Teaching Biostatistics in Medicine and Allied Health Sciences

Damian J. J. Farnell • Renata Medeiros Mirra

Editors

Teaching Biostatistics in Medicine and Allied Health Sciences

Editors
Damian J. J. Farnell
School of Dentistry
Cardiff University
Cardiff, UK

Renata Medeiros Mirra
School of Dentistry
Cardiff University
Cardiff, UK

ISBN 978-3-031-26012-4 ISBN 978-3-031-26010-0 (eBook)
https://doi.org/10.1007/978-3-031-26010-0

This Springer imprint is published by the registered company Springer Nature Switzerland AG
The registered company address is: Gewerbestrasse 11, 6330 Cham, Switzerland

Preface

There are many textbooks that relate to the theory and practical uses of biostatistics in medicine and dentistry. However, there are few (if any) that relate to the theory and practice of *teaching* biostatistics to undergraduate or postgraduate students. This book aims to address this "gap in the market" by compiling articles from the "Burwalls" conferences in 2020 and 2021 in order to provide personal insights and first-hand experiences of biostatistics teaching. Burwalls is an annual conference for people involved in teaching statistics and evidence-based medicine to students or professionals in medicine, health and social care in Higher Education (HE), the NHS, or similar institutions in the UK. This conference began in 1980 in a grand house called Burwalls in Bristol and it continued in this manner for quite some time. Eventually though, the Burwalls meeting became "peripatetic", i.e. it was (and still is) held each year at institutions in the UK where participants work. Burwalls is a friendly conference with the aims of encouraging and equipping professionals with the tools they need to teach biostatistics, giving people a chance to network and promoting collaborations, and allowing them to share experiences of teaching.

We remark that the COVID-19 pandemic had a profound effect on HE teaching during 2020 and 2021 and that this was reflected strongly in the content presented in the Burwalls conferences in these 2 years. A short 1-day (online) meeting occurred in 2020 due to the pandemic, whereas a full 3-day (online) conference occurred in 2021. The meeting in 2021 was organised broadly into topics on learning and teaching, online learning (and the effects of the COVID-19 pandemic), and specific software tools and resources. Contributions to this volume have also been broadly organised along these lines. We have tried to capture discussions that occurred at Burwalls in 2020 and 2021 in this book. Although it was not possible to include all of them in this volume, many of the chapters were written with these discussions in mind. We remark also that biostatistics teaching comes with its own specific set of challenges. For example, one is often teaching complex mathematical or statistical ideas to people whose primary training is not in these fields. This can lead to a fear of the topic or a feeling that it is not relevant. Hopefully, this volume will also give the reader an idea of how to overcome some of these issues.

Chapter "A Survey of Biostatistics Teaching in Medicine and Dentistry in Higher Education in the UK" presents a description of the nature and extent of biostatistics teaching occurring in HE in the UK and it provides good context for the rest of this volume. Chapters "Evidence-Based Practice Teaching for Undergraduate Dental Students", "Teaching Medical Statistics Within the Context of Evidence Based Medicine", "Teaching Null Hypothesis Significance Testing (NHST) in the Health Sciences: The Significance of Significance", "Teaching Conceptual Understanding of p-Values and of Confidence Intervals, Whilst Steering Away from Common Misinterpretations", "Using Directed Acyclic Graphs (DAGs) to Represent the Data Generating Mechanisms of Disease and Healthcare Pathways: A Guide for Educators, Students, Practitioners and Researchers"; and "Statistics Without Maths: Using Random Sampling to Teach Hypothesis Testing" might be placed very broadly under the banner of "teaching and learning", albeit on topics as diverse as teaching evidence-based practice, the pitfalls of null-hypothesis significance testing, Directed Acyclic Graphs (DAGs), and even how one can go about teaching statistics without mathematics. Online learning is a theme that has become increasingly common at the Burwalls meetings over recent years, and this topic is explored in chapters "COVID-19: Online Not Distant—MSc Students' Feedback on an Alternative Approach to Teaching 'Research Methods and Introduction to Statistics' at UCL Queen Square Institute of Neurology" and "Common Misconceptions of Online Statistics Teaching". Clearly, this also reflects changes in teaching in HE more generally, especially in light of the impact of the COVID-19 pandemic in 2020 and 2021. A common comment from participants at Burwalls at this time was that online teaching has many advantages, as well as disadvantages, and this is reflected in these chapters also. The topic of online learning also leads on naturally to discussions in chapters "Authentic Project-Based Assessment Using the *Islands*: Instructor's View", "An Interactive Application Demonstrating Frequentist and Bayesian Inferential Frameworks", "Teaching Data Analysis to Life Scientists Using "R" Statistical Software: Challenges, Opportunities, and Effective Methods" of specific software tools that can be used to teach biostatistics, such as a virtual web-based population on an island, an interactive online "app" that demonstrates frequentist and Bayesian inferential approaches, and how one can use the software package "R" to provide an effective teaching platform. It is clear that the modern-day teacher of biostatistics needs to know how to use such online and/or software tools and resources. At the very least, they form valuable teaching resources that can provide entertaining, novel, and effective tools. We complete this volume with discussions of more general issues relating to biostatistics teaching that have taken placed during the Burwalls conferences over the years. The boundary between science and statistics is explored in chapter "Statistics in a World Without Science". We then move on to a description of discussions during the 2021 meeting in chapter "Killing Me Softly with Your Stats Teaching: How Much Stats Is Too Much Stats?" relating to the depth of the statistics that we ought to be teaching our students, i.e. "how much stats is too much stats?" Issues relating to career progression as a statistician working in academia in the UK and also statistical consultancy as part of a "life in biostatistics" are considered in the final chapter (chapter "Life as a Medical Statistician").

Finally, we wish to thank all participants and speakers at Burwalls 2020 and 2021 for their valuable contributions to these conferences. In particular, we would like to thank the authors for all of their hard work and for providing such an excellent, interesting, and diverse selection of papers.

Cardiff, UK   Damian J. J. Farnell
Cardiff, UK   Renata Medeiros Mirra
30 September 2022

Contents

A Survey of Biostatistics Teaching in Medicine and Dentistry in Higher Education in the UK

Damian J. J. Farnell

1   Introduction

The importance of biostatistics teaching is being increasing recognised in medical education, e.g. in the context [1] of "Big Data" (i.e. extremely large or complex data sets). Despite this, research into biostatistics education is a relatively new field, although it is developing strongly [2, 3]. Much of this previous work has focussed on the specific problems or challenges (and solutions) of teaching biostatistics (see, e.g. [4, 5]). However, research into statistics pedagogy is [2] "disconnected, fragmented, and difficult to access" (see also Ref. [3]). This statement mirrors the varied nature of biostatistics teaching within the UK. Indeed, it is fair to say that there is only limited guidance from professional clinical bodies in the UK as to the depth and amount of biostatistics that students should learn. Here, I provide an overview of biostatistics teaching in medicine and dentistry in higher education in the UK.

2   Materials and Methods

An invitation is generally sent out to all participants of the Burwalls meetings (for teachers of biostatistics) each year to complete a survey of biostatistics teaching within their respective schools at UK universities. Twenty documents [6] were collected (mostly in 2019) from medical and dental schools across the UK, namely Cardiff, Hull & York Medical School, Kings College London, Newcastle, Queen's University Belfast, St George's London, University College London, Aberdeen, Birmingham, Bristol, East Anglia, Edinburgh, Exeter, Leicester, Nottingham, Oxford, and Sheffield. Data and analyses given here formed the basis of an oral presentation by the author to the Burwalls meeting in 2020.

D. J. J. Farnell (✉)
School of Dentistry, Cardiff University, Cardiff, UK
e-mail: farnelld@cardiff.ac.uk

Quantitative data collected in this survey included the academic years in which biostatistics is taught, the number of students per years, estimated total hours of teaching, software used, recommended textbooks, and course content. Note that respondents were able to "tick" items on a list of concepts and (separately) a list of calculations or equations that are introduced during their biostatistics teaching. Respondents were also asked to provide an overview of teaching of biostatistics at their institutions as free-text comments (i.e. qualitative data). They were also asked to describe details of assessments as free-text comments. Qualitative analysis of this free-text data was carried out by using NVivo V12. Common keywords or expressions were identified in this free-text data; this led to the identification of various broad themes relating to delivery of teaching, course content, assessment types, and motivation for/focus of the teaching. Keywords or expressions were then "coded" using NVivo for each document and the frequency of occurrence of these keywords or expressions with respect to the 20 documents was found. Results are illustrated graphically as "word clouds" below.

3   Results

Results for typical class sizes varied from a handful of students to 500 students (median = 203; IQR = 145), whereas estimated teaching hours per year varied from 2 to 40 h (median = 15; IQR = 16.5). Teaching of biostatistics occurred in just one year of an undergraduate medical degree for the majority of programmes of learning (first or second year generally), although it could occur over all five years in some cases and/or at postgraduate level. Free-text comments showed that the most common forms of delivery of biostatistics courses were lectures, online or e-learning, and tutorials, as shown in Fig. 1. However, other methods of teaching were used also practicals—software, research consultation or support, flipped learning,

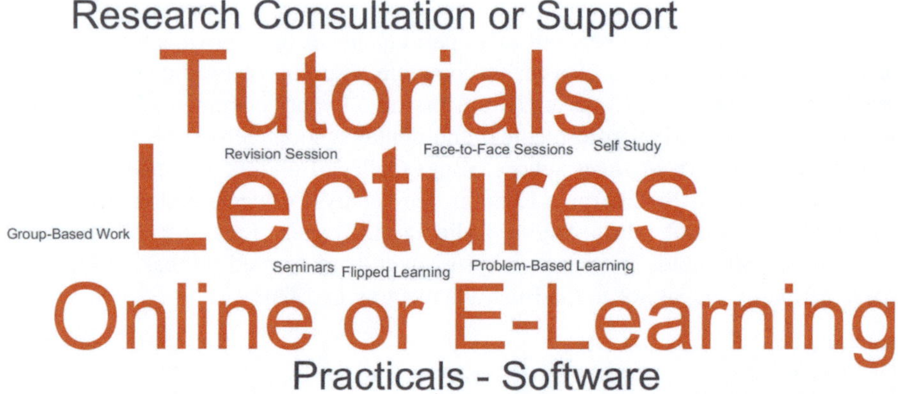

Fig. 1 Word cloud of methods keywords or expressions in free-text comments for the theme: delivery of teaching

group-based work, problem-based learning, revision sessions, and self-study. Indeed, two or more delivery types were often reported as being used in concert during a course or programme of learning.

Results for core concepts and (separately) the types of calculations (including equations) taught are shown in Table 1. The mean number of concepts taught per course or programme of learning was 12.4 (median = 14), whereas the number of calculations was much lower (mean = 4.8; median = 1). These results show that teaching of theoretical calculations or equations in biostatistics was less common in the UK than teaching of basic concepts. This difference is shown to some extent in free-text comments, as shown in Fig. 2. We see that the most frequent keywords or expressions are statistics, epidemiology, study design, and critical appraisal. However, it is noticeable that "presentation of results" and "interpretation of results" were mentioned also, as well as "NOT theory or formulae" and "no analytical element". Indeed, a direct quote along these lines from these free-text comments was that: *"The focus of the course is on interpretation. At the end of the year students are expected to be able to interpret common results tables - which you might find in a*

Table 1 Concepts and calculations taught (percentages with respect to the 20 documents)

	Primary	Secondary	Other (reported in 1 document only, i.e. 5%)
Concepts	Confidence intervals (95%), hypothesis testing and P-values (90%), comparing two means (90%), concepts of populations and samples (90%), types of variables (90%), distributions (85%), summary statistics (85%), critical appraisal (80%), graphs (75%), comparing two proportions (70%), linear regression (60%)	Logistic regression (40%), survival analysis (35%), multivariate analysis (30%), odds and risk ratios (25%), ANOVA (20%), meta analyses (20%), non-parametric tests (20%), sensitivity/specificity (15%), chi-squared analysis (10%), diagnostic test accuracy (10%), probability (10%), randomised trials (10%), study design (10%)	Agreement (e.g. kappa), bias, bootstrapping, causality, communication of risk, confounding, correlation coefficients, history of biostatistics, internal/external validity, measurement reliability, observational study, prevalence or incidence, sample sizes or power calculations, z-scores
Calculations	Confidence intervals (55%), comparing two means (45%), hypothesis testing and P-values (45%), comparing two proportions (25%), summary statistics (25%), populations and samples (25%), distributions (25%), linear regression (25%)	ANOVA (15%), non-parametric tests (15%), odds & risk ratio (15%), chi-squared analysis (15%), correlations (10%), logistic regression (10%), survival analysis (10%)	Diagnostic test accuracy, longitudinal analyses, meta-analysis, multivariate analysis, probability, sample size estimation, z-scores

Interpretation of Results
Data Analysis or Science
Critical Appraisal
Qualitative Methods Advanced Statistical Techniques Application to Patients
Research Ethics Concepts & Theory
Statistics
Literature Searches
Systematic Reviews
Communication & Documentation
NOT Theory or Formulae No Analytical Element
Epidemiology
Study Design
Presentation of Research or Statistical Results

Fig. 2 Word cloud of methods keywords or expressions in free-text comments for the theme: course contents

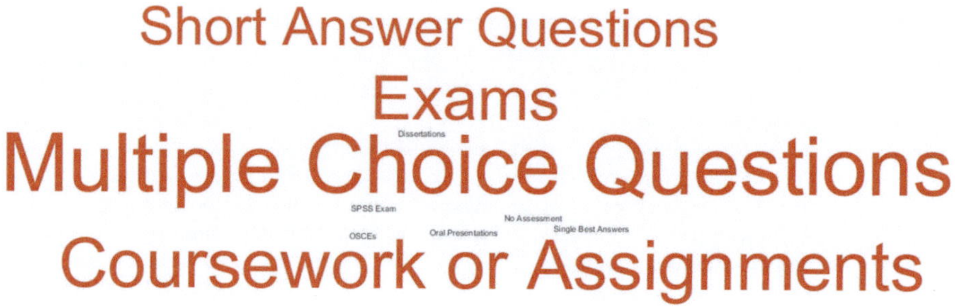

Short Answer Questions
Exams
Dissertations
Multiple Choice Questions
SPSS Exam
No Assessment
OSCEs Oral Presentations Single Best Answers
Coursework or Assignments

Fig. 3 Word cloud of methods keywords or expressions in free-text comments for the theme: types of assessment

BMJ publication for example. There is no expectation that they can undertake any analysis."

Interestingly, a majority of documents (i.e. 11 out of 20 or 55%) reported that the use of statistical software was not taught at all. Where it was taught, the most common software packages were reported as being SPSS (4 out of 20 documents or 20%), followed by Minitab (10%), STATA (10%), GraphPad Prism (5%), JASP (5%), and R (5%). Assessment methods were similarly varied, as illustrated in Fig. 3 for free-text comments. These were, namely, multiple choice questions, coursework or assignments, exams, short answer questions dissertation, oral presentations, Objective Structured Clinical Examinations (OSCEs), "single best answers" questions, and software exams (SPSS). A wide range of textbooks [7–26] were recommended by respondents; the most commonly used textbooks mentioned in these documents were given by: Bland (2015) (4 documents; Ref. [8]), Kirkwood and Sterne (2010) (6 documents; Ref. [14]), Petrie and Sabin (2019) (4 documents; Ref. [18]), and Campbell, Machin, and Walters (2010) (3 documents; Ref. [15]).

Free-text comments in these documents indicated that there were three main focusses of/motivations for carrying out biostatistics teaching, namely research (mentioned in 9 documents), evidence-based medicine (6 documents), and public health/epidemiology (2 documents).

4   Conclusion

The results presented in this chapter show that a wide range of biostatistics teaching occurred in higher education in the UK up to 2020; it is likely that very little has changed at the time of publication of this volume. This variability applied to all of the fields considered here, including class sizes, numbers of years over which bio-statistics was taught, methods of delivery of teaching, types of assessments, software employed (if at all), and even in the motivations for teaching (etc.). However, a common "core" of concepts and types of calculations (including equations) does emerge from the data, as shown in Table 1. Although these lists should come as no surprise to anyone who has taught biostatistics (indeed, Table 1 agrees with sylla-buses quoted in the literature, e.g. Ref. [27–30]), it is still very useful to have them written down explicitly in Table 1. This table should therefore provide a useful resource for anyone interested in biostatistics teaching, as well as providing excellent context for the rest of this book. It is noticeable that there was a stronger emphasis on the "practical" aspects of biostatistics rather than the more "theoretical" aspects of this discipline; this is understandable given that typical cohorts are medical, dental, and other "allied health" students. Finally, biostatistics plays an important role in the analysis of research data; this came through clearly in free-text comments provided by respondents as a primary "motivator" for teaching biostatistics. Furthermore, understanding of statistical concepts is an important part of evidence-based medicine, as well as in public health/epidemiology. This was also reflected as "motivators" in these free-text comments. Again, biostatistics plays a key role in medicine and dentistry (and beyond); it therefore needs to be taught effectively and it needs to be taught well.

Acknowledgements I wish to acknowledge and thank Dr. Margaret MacDougall of Edinburgh University for collecting and maintaining the survey of participants of Burwalls each year. (The documents used as raw data for this chapter are held at Ref. [6].) I wish to thank Dr. Renata Medeiros Mirra for her helpful and insightful comments and feedback relating to this paper.

References

1. Brimacombe MB. Biostatistical and medical statistics graduate education. BMC Med Educ. 2014;14(1):1–5.
2. Tishkovskaya S, Lancaster GA. Statistical education in the 21st century: a review of challenges, teaching innovations and strategies for reform. J Stat Educ. 2012;20(2):1–20.
3. Zieffler A, Garfield J, Alt S, Dupuis D, Holleque K, Chang B.. What does research suggest about the teaching and learning of introductory statistics at the college level? A review of the literature. J Stat Educ. 2008;16(2).

4. Sahai H. Teaching biostatistics to medical students and professionals: problems and solutions. Int J Math Educ Sci Technol. 1999;30(2):187–96.

5. Ojeda HSM. Problems and challenges of teaching biostatistics to medical students and professionals. Med Teach. 1999;21(3):286–8.

6. www.ed.ac.uk/usher/annual-meeting-teachers-of-medical-statistics-2018/overview-of-teaching-of-statistics-within-medicine. Accessed 6-10-2022.

7. Donovan D, McDowell I, Hunter D. AFMC primer on population health: a virtual textbook on public health concepts for clinicians. Ottawa: AFMC Association of Faculties of Medicine of Canada; 2013.

8. Bland M. An introduction to medical statistics. Oxford: Oxford University Press; 2015.

9. Goldacre B. Bad science. London: Fourth Estate; 2009.

10. Harris M, Taylor G, Taylor J. Catch up Maths & Stats: for the life and medical sciences. Oxfordshire: Scion; 2005.

11. Smeeton N. Dental statistics made easy. Abingdon: Radcliffe Medical Press; 2005.

12. Field A. Discovering statistics using IBM SPSS statistics. Thousand Oaks: Sage; 2013.

13. Webb P, Bain C, Pirozzo S. Essential epidemiology: an introduction for students and health professionals. Cambridge: Cambridge University Press; 2005.

14. Kirkwood BR, Sterne JA. Essential medical statistics. Chichester: Wiley; 2010.

15. Campbell MJ, Machin D, Walters SJ. Medical statistics: a textbook for the health sciences. Chichester: Wiley; 2010.

16. Campbell MJ, Machin D. Medical statistics. A commonsense approach. Chichester: Wiley; 1999.

17. Peat J, Barton B. Medical statistics: a guide to data analysis and critical appraisal. Chichester: Wiley; 2008.

18. Petrie A, Sabin C. Medical statistics at a glance. Chichester: Wiley; 2019.

19. Harris M, Taylor G. Medical statistics made easy. Boca Raton: CRC Press; 2003.

20. Altman DG. Practical statistics for medical research. Boca Raton: CRC Press; 1990.

21. Peacock JL, Kerry SM, Balise RR. Presenting medical statistics from proposal to publication. Oxford: Oxford University Press; 2017.

22. Davis C. SPSS step by step: essentials for social and political science. Chicago: Policy Press; 2013.

23. Pallant J. SPSS survival manual: a step by step guide to data analysis using IBM SPSS. London: Routledge; 2020.

24. Swinscow TDV, Campbell MJ. Statistics at square one. Chichester: Wiley; 2002.

25. Machin D, Bryant T, Altman D, Gardner M. Statistics with confidence: confidence intervals and statistical guidelines. Chichester: Wiley; 2013.

26. Peacock J, Peacock P. Oxford handbook of medical statistics. Oxford: Oxford University Press; 2011.

27. Sami W. Biostatistics education for undergraduate medical students. Biomedica. 2010;26(1):80–4.

28. Windish DM. Brief curriculum to teach residents study design and biostatistics. BMJ Evid Based Med. 2011;16(4):100–4.

29. Harraway JA, Sharples KJ. A first course in biostatistics for health sciences students. Int J Math Educ Sci Technol. 2001;32(6):873–86.

30. Ambrosius WT, Manatunga AK. Intensive short courses in biostatistics for fellows and physicians. Stat Med. 2002;21(18):2739–56.

Evidence-Based Practice Teaching for Undergraduate Dental Students

Sam Leary and Amy Davies

1   Introduction

Evidence-based practice (EBP) was originally defined as an integration of clinical expertise with published clinical evidence, to inform decisions about the care of individual patients [1]. Applying EBP ensures that clinicians provide the most effective care available for their patients. Evidence-based medicine was formally recognised as a way of teaching medicine by 1992 [2], followed by the establishment of evidence-based dentistry (EBD) in 1995 [3]. However, the majority of dentists, although aware of EBD, do not apply it in practice; barriers to implementation include a lack of time and inadequate training [4].

The UK's General Dental Council (GDC) guidelines "Preparing for Practice" which were published in 2011 and updated in 2015 [5] include intended learning outcomes for undergraduate dental students which focus on EBD, critical appraisal, and epidemiology (§1.1.1, §1.1.2 and §1.1.12), but do not specifically mention statistics and data analysis. Interpretation of these guidelines is likely to vary across the 16 UK dental schools, and the amount of statistics included in undergraduate dental courses, along with whether or not statistics is taught separately from EBD/P is not well documented. However there is some information on dentistry courses, in particular regarding which statistical concepts are taught, included in an overview of the undergraduate teaching of statistics within medicine and allied health sciences across UK universities, which is available online [6].

Bristol Dental School (University of Bristol, UK) introduced a new curriculum for the 5-year Bachelor of Dental Surgery (BDS) degree in the academic year 2019–2020, known as BDS21 (dentistry in the twenty-first century). The previous curriculum, known as BDS18 (the final cohort started in 2018), had run for 12 years, and included a theme of Oral Health Research (OHR). This theme started in year 2

S. Leary (✉) · A. Davies
Bristol Dental School, University of Bristol, Bristol, UK
e-mail: s.d.leary@bristol.ac.uk

© The Author(s), under exclusive license to Springer Nature Switzerland AG 2023
D. J. J. Farnell, R. Medeiros Mirra (eds.), *Teaching Biostatistics in Medicine and Allied Health Sciences*, https://doi.org/10.1007/978-3-031-26010-0_2

with the Quantitative Research Methods (QRM) course which was delivered using a flipped classroom approach (e-lectures and small group tutorials). The QRM course covered basic epidemiological concepts, different types of study design, summarising and interpreting data, and choosing appropriate statistical analyses, with clinical examples from research publications used to illustrate concepts; a written exam was used for summative assessment. Full details of the development of this course are provided in Leary and Ness [7]. The students then completed a Critical Appraisal Project (CAP) in year 3, where they firstly appraised a published paper on a specific oral health topic in a small tutor-led group, then individually critically appraised another paper on the same topic, produced a written report, and gave an oral presentation for summative assessment. The final component of the theme was an Evidence Summary Project (ESP) in year 4, where students worked in pairs to produce a comprehensive review of 7–10 published papers on a clinical question of their choice. Each pair was assigned a mentor and wrote a report which was used for summative assessment.

In preparation for BDS21, a rigorous review of BDS18 by senior staff within Bristol Dental School was undertaken. The overall objective of this new curriculum is to provide an integrated, helical (i.e. aspects running across multiple years of the degree) and sustainable curriculum that is contemporary, evidence-based, and supports the students in reaching their full potential. The theme of OHR was replaced by the helical theme of EBP, again based on the GDC guidelines. Within this, the QRM course was modified slightly and moved to year 1 (to allow reduction of content in the previously overcrowded year 2), the CAP project was replaced by a series of critical appraisal workshops in years 2 and 3 (to ensure that all students have as similar a learning experience as possible and also reduce tutor workload), and the ESP remained in year 4. Each component of the EBP theme is described in Sect. 2 of this chapter, and the experience of delivering this theme is summarised in Sect. 3.

2   Description of the EBP Theme Components

The structure of the EBP theme is illustrated in Fig. 1.

BDS21 is being introduced 1 year at a time; hence year 1 is now running for the third time, year 2 for the second time, year 3 for the first time, and year 4 will run for the first time in 2022 (although the ESP was part of BDS18 so is already well-established). All the teaching materials for all years of the theme are available to the students on the University of Bristol Blackboard virtual learning environment (Blackboard Learn, Washington, US).

The EBP theme focuses on concepts and interpretation. This is because most students will not need to understand statistical theory, learn formulae, or perform calculations in their future careers. Excluding these is also likely to minimise "statistical anxiety" [8]. Statistics packages are not taught, so as not to detract from improving understanding of the underlying concepts [9].

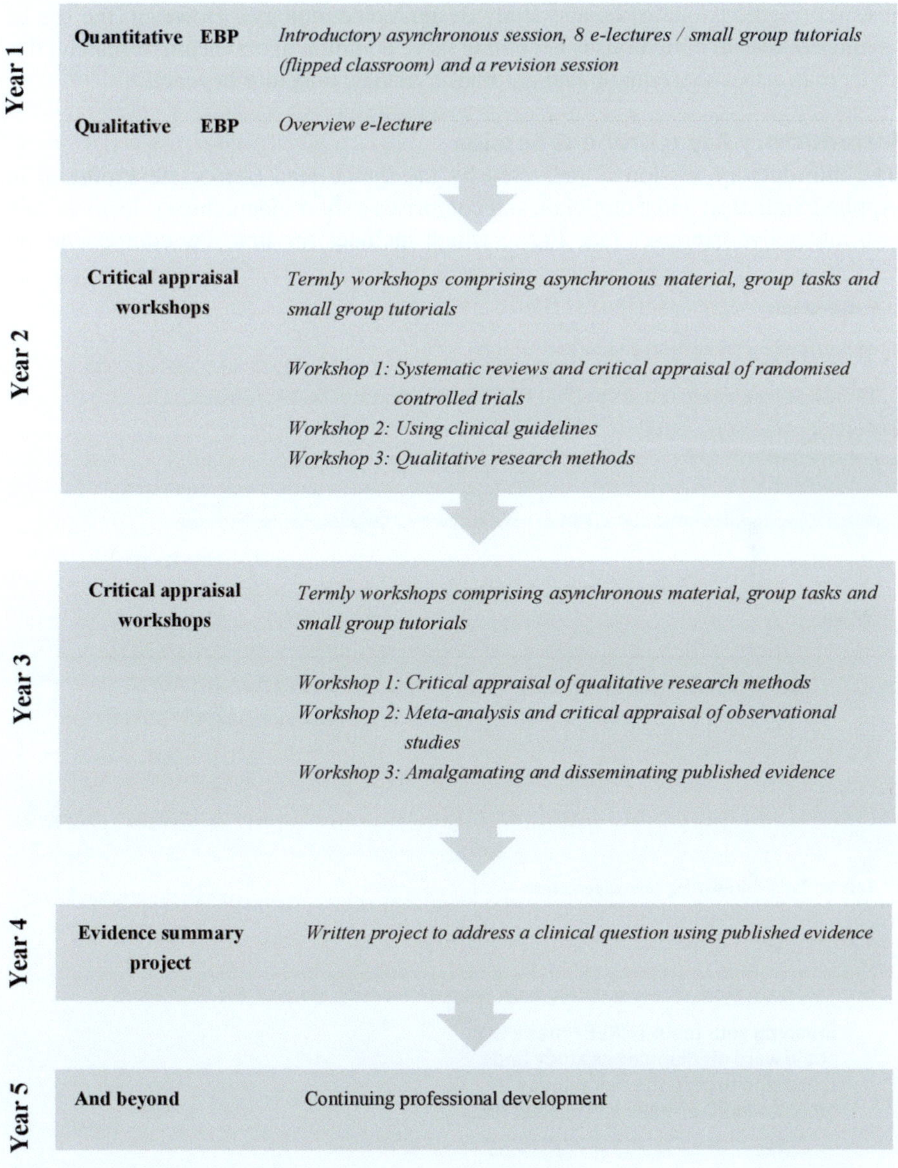

Fig. 1 Structure of the evidence-based practice (EBP) theme

2.1 Year 1 EBP

Year 1 EBP comprises an asynchronous (i.e. does not have to be undertaken at a specific time) introductory session, eight e-lectures with associated small group tutorials, and a revision session. The main emphasis of the EBP theme is quantitative research methods (statistics and study design/epidemiology). However, the use of qualitative research methods is becoming more common in oral health research [10], so there is also a short qualitative methods overview e-lecture in year 1.

Introductory Asynchronous Session

The introductory session is delivered by the theme lead (Associate Professor in Applied Statistics) and a clinician, and comprises a short video, three e-lectures, and two interactive exercises (see Fig. 2 which includes the first few components on

Session details

Welcome to the theme of Evidence Based Practice (EBP).

This is an asynchronous session timetabled for 6th October 2021, with the following intended learning objectives:

- Appreciate the importance of EBP
- Understand the structure of the year 1 EBP sessions
- Be aware of future career options available to dentists

Please start by watching the short video below, which will explain the structure of this session, then click "mark reviewed".

EBP Intro

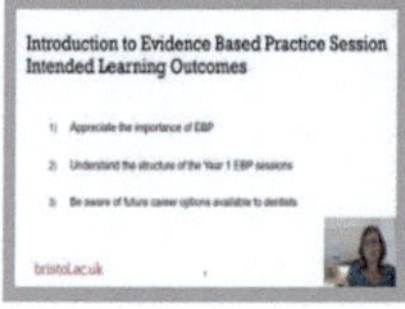

Task 1a: Patient presenting newspaper article

If a patient arrives for their dental appointment with the following newspaper article published in the Mail Online, and asks if brushing their teeth more frequently will prevent them getting diabetes, think about how you would react (click on the link to read the full article):

Brushing your teeth THREE times a day could ward off diabetes as study finds those who exceed the twice-a-day recommendation have a 14% lower risk

- A study looked at the oral hygiene of almost 190,000 people from South Korea
- They were followed for an average of 10 years to see who developed diabetes
- Those who brushed three times a day had a 14 per cent lower risk of diabetes

By VANESSA CHALMERS HEALTH REPORTER FOR MAILONLINE
PUBLISHED: 23:01, 2 March 2020 | UPDATED: 23:20, 2 March 2020

https://www.dailymail.co.uk/health/article-8066597/Brushing-teeth-THREE-times-day-ward-diabetes-study-finds-14-lower-risk.html

Reflect on how you would react in the padlet below, then click "mark reviewed".

Dr Leary will summarise the responses to this task, and suggest some issues to consider after the 8th October 2021; this document will then be available on Blackboard.

https://padlet.com/sdleary/ucvldiwqq0yxwk92

Fig. 2 Excerpt from year 1 evidence-based practice introductory asynchronous material

Blackboard). The main purpose of this session is to explain the structure of the year 1 EBP course and how it is built upon in future years of the dental degree, and describe the importance of EBP in dentistry.

E-Lectures and Small Group Tutorials

The flipped classroom approach [11] is used for the majority of the year 1 teaching. Didactic teaching is in the form of eight e-lectures each lasting 20–25 min produced by the theme lead, split into chapters to allow easy navigation, with pop-up multiple-choice questions to engage students. Small group interactive, structured tutorials (one per e-lecture) are used to reinforce the information covered in the e-lecture, hence do not include any new material; tutors are quantitative staff from Bristol Dental School. Oral health examples from research publications are used to illustrate concepts in both the e-lectures and tutorials. Table 1 shows the topics and content included in the eight sessions.

The e-lectures were created using Windows Mediasite Recorder (Sonic Foundry Inc, Wisconsin, US), although these will be re-recorded using narration within PowerPoint, as this allows specific slides to be updated in isolation. Quiz questions are incorporated in the e-lectures within Blackboard. E-lecture viewing is timetabled, but students can choose to watch the e-lectures at the time of day that suits them best for studying, as long as it is before the associated tutorial, and can also rewatch them for revision.

Each e-lecture is followed by a 2-h compulsory small group (one sixth of the year) tutorial; each group of students is assigned a statistically trained tutor for the whole course, to ensure as much consistency as possible. Pre-prepared materials are provided for the tutors, to minimise the impact of different groups having different tutors, and also the workload for the tutors. Each tutorial consists of an introductory exercise followed by a series of problems for the students to work through, plus plenty of opportunity to ask questions. Introductory exercises are either e-voting quizzes (anonymous voting via electronic devices using TurningPoint, Turning

Table 1 Topics and content of year 1 quantitative evidence-based practice sessions

Topic	Content
Introduction to study design	Basic epidemiological concepts, common types of study design, hierarchy of evidence, causal associations
Introduction to summarising data	Types of variables, graphical presentation, prevalence and incidence, measures of location and variability
Randomised controlled trials	Definition, planning, conducting, analysing, strengths and weaknesses
Understanding statistical inference	Sampling/inference, accuracy and precision, confidence intervals, p values
Cohort studies	Definition, examples, risk ratios, strengths and weaknesses
Investigating hypotheses	Purpose of analysis, comparing numerical/categorical outcomes between groups, assessing agreement between numerical/categorical outcomes
Case-control studies	Definition, example, odds ratios, strengths and weaknesses
Assessing associations	Correlation, linear regression, other types of regression modelling, adjusting for confounders

Technologies, Ohio, US) or group work where students are asked to design a study to address a given hypothesis. In the problem-based part of the tutorial students are provided with a brief summary of a published oral health example. They work through a series of short answer questions on the key concepts in pairs or small groups, then the answers are discussed as a whole group. The e-lectures/tutorials are spread over an approximately 6-month period, and all students are required to provide a recap of one of the sessions at some point during the course, which is used as an indicator of engagement.

Summative Assessment

At the end of the course all students have a revision session in their small groups to practice mock exam questions, and a glossary of terms used in the course is also provided. EBP questions are included in the end of year programme-based assessment. Six questions are included in the 120-question single best answer section (SBA) and two 10-mark questions are included in the 140-mark multiple short answer (MSA) section; this allocation is based on teaching time for EBP relative to other themes.

2.2   Years 2 and 3 EBP

Years 2 and 3 comprise six compulsory workshops, one per term, each including asynchronous material (maximum 3-h), student-led group work to complete a task (timetabled for 3-h when required), and tutorials (1.5 h, between one sixth and half of the year at a time). The workshops are intended to bridge the gap between the EBP knowledge gained in year 1 and the requirement to produce a written ESP in year 4. Students are assessed formatively through the tasks, to demonstrate engagement with the theme. The workshops are delivered by the theme lead, quantitative and qualitative staff, and clinicians from Bristol Dental School. The first workshop is described below, along with brief summaries of the other five workshops.

Systematic Reviews and Critical Appraisal of Randomised Controlled Trials (RCTs) (Year 2 Term 1)

The asynchronous material for this workshop comprises a very short introductory video, recap quiz and recap e-lecture for relevant year 1 EBP topics, overview of systematic reviews e-lecture with an associated activity based on searching the literature, video to introduce the clinical relevance of the workshop, overview of a specific systematic review paper e-lecture which includes pop-up questions [12], critical appraisal checklist example e-lecture (using the Critical Appraisal Skills Programme (CASP)) [13] tool for RCTs and a summary quiz; see Fig. 3 for an example of some of the Blackboard material.

The students are divided into groups of approximately four students. Each group is required to apply the CASP tool to a specific RCT and create slides to present their findings during their tutorial. Within each tutorial, three different RCT papers are presented by the students; there is opportunity for discussion with questions from peers encouraged, then the tutor provides verbal feedback.

Section 1: Introduction

The theme of Evidence Based Practise (EBP) which begun in year 1 will be continued in year 2 through a series of workshops.

The first workshop on Systematic Reviews and Critical Appraisal of RCTs and begins with an asynchronous session timetabled for 17th November 2021, followed by live small group tutorials on Wednesday 24th November 2021.

The intended learning objectives are:

1. Define a systematic review
2. Apply a critical appraisal tool to randomised controlled trials
3. Summarise the quality of a randomised controlled trial

Note the key difference between a systematic review and a critical appraisal is that <u>a systematic review is based on multiple papers, and a critical appraisal is based on a single paper</u>.

Systematic Reviews and Critical Appraisal of RCTs - Introduction

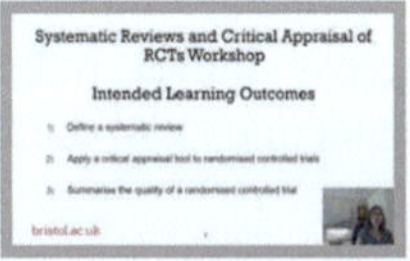

Section 2: Recap quiz based on Year 1 EBP sessions

Have a go at this quiz to see what you can remember from the eight year 1 EBP sessions. Press submit and then click OK to review your responses and the correct answers. Once an attempt is submitted the next section of this material will automatically become accessible.

Fig. 3 Excerpt from year 2 evidence-based practice workshop material

Using Clinical Guidelines (Year 2 Term 2)

This workshop starts with introductory and overview e-lectures, then students are divided into groups to prepare a patient information leaflet to highlight the key UK guidelines on an assigned topic, e.g. tooth whitening. The students are also required to review leaflets on two other topics, update their own leaflet based on peer feedback, then present it orally during a tutorial.

Qualitative Research Methods (Year 2 Term 3)

The asynchronous material for this workshop consists of e-lectures covering different aspects of qualitative research methods, plus activities based on journal articles and videos which require completion of self-directed learning forms. Tutorials include discussion of these forms, and also qualitative interview exercises in small groups.

Critical Appraisal of Qualitative Research Methods (Year 3 Term 1)

For this workshop the students start by watching an introductory e-lecture, then individually apply the CASP tool for qualitative studies to two published papers. Students are provided with tutor feedback on their checklists during their tutorials.

Meta-analysis and Critical Appraisal of Observational Studies (Year 3 Term 2)

The format for the asynchronous part of this workshop is very similar to the first workshop in year 2, and comprises videos, e-lectures, quizzes, and an individual activity based on interpreting a meta-analysis. The group work task requires students to apply a CASP tool to a specific cohort or case-control study, then produce two slides which summarise the strengths and limitations of that study. During their tutorial the students present their slides, and the tutor provides verbal feedback.

Amalgamating and Disseminating Published Evidence (Year 3 Term 3)

This workshop will be delivered for the first time in June 2022 so the materials are not yet developed, but the intended learning objectives are to synthesise evidence across studies, judge the relative quality of studies, and develop skills in writing for academics and the lay public.

2.3   Year 4 EBP

The final component of the EBP theme is the ESP, which will run for the first time within BDS21 from September 2022, but the format will be the same as the BDS18 version of the course. Students are introduced to the project via e-lectures which give overviews of the approach to the project, literature searching, and writing the 4000-word report. Additional literature searching resources and a report template are provided, and there is a timetabled synchronous session given by the ESP lead (Senior Lecturer) and Dental School librarian which provides further clarity of requirements and an opportunity to ask questions. Students work in pairs, with guidance from a mentor (member of the Dental School staff), to produce a comprehensive review of 7–10 published papers on a clinical question of their choice in a 4000-word report; see Table 2 for some of the recent questions that were addressed.

Students firstly submit their title, PICO (Population/, Intervention/Control/ Outcome) question, written methods section, search strategy, and flow diagram for formative feedback from the Dental School librarian. They then submit their full report for formative feedback from their mentor. After responding to this feedback,

Table 2 Examples of clinical questions addressed in year 4 evidence-based practice projects

In healthy individuals with no systemic illnesses, how does body mass index affect the relative abundance of bacterial species in saliva?
How effective is the use of glass ionomer cement, as compared to use of resin-based fissure sealant in preventing dental caries in the molars of children?
Following a dental extraction, is the incidence of dry socket increased in adult tobacco smokers in comparison to adult non-smokers or ex-smokers?
How does chewing sugar-free or sugar-substituted gum after meals affect the development of caries in the permanent dentition?
In adults, are electric toothbrushes more effective in improving clinical parameters of periodontal health than manual toothbrushes?

they submit the final version of their report for summative assessment; these reports are double marked by their mentor and another member of staff (one clinical, one non-clinical).

3   Experience of Running the EBP Theme

The results and feedback relate to the 2020–2021 academic year where years 1 and 2 were part of the BDS21 curriculum and year 4 was part of the BDS18 curriculum. There were 70–80 students per year, and the majority attended most compulsory EBP sessions, although some sessions were missed due to COVID-19-related reasons.

3.1   Assessment Results

The year 1 summative assessment was moved online due to COVID-19, and therefore the SBA questions were removed. For the two MSA questions on EBP combined, the mean (standard deviation) score was 68.2% (16.6%); this was the second highest mean value across all nine themes included in this programme-based assessment.

For year 2, all students engaged with all three EBP workshops satisfactorily, by completing the required tasks. For year 4, four students (two projects) failed the EBP summative assessment but passed after revision of their projects based on feedback; the mean (standard deviation) project mark was 63.9% (8.7%).

3.2   Feedback

Student Feedback

Students were asked to complete a series of multiple-choice (using 5-point Likert scale) and free text response questions for each subject in each year of BDS21 during a synchronous session; the response rates were variable (42% for year 1, 76%, 72%, and 30% respectively for the year 2 workshops).

Despite several of the year 1 EBP tutorials needing to be delivered online via Blackboard Collaborate due to COVID-19 restrictions, student feedback was generally very positive. The key strengths identified were the clear and informative e-lectures with pop-up questions, and also the small group interactive tutorials which were felt to consolidate learning in an enjoyable and interactive way. EBP was perceived to be a difficult subject by some, but the e-lectures and tutorials to aid understanding were appreciated, and no areas for improvement were identified. Table 3 presents the responses from the multiple-choice questions.

For the first year 2 EBP workshop, there were many positive comments on the 3-h asynchronous session (e.g. well-structured, good variety of activities, helpful to recap relevant year 1 EBP topics), and the group work task was generally very well

Table 3 Year 1 student feedback for evidence-based practice

Question	% completely/somewhat agree[a]
Staff have made evidence-based practice interesting	74
The teaching helped me to learn about evidence-based practice	94
I understood what was required of me and why	88
Evidence-based practice helped me in my developmental journey to becoming a qualified dental professional	70
The main resources used for evidence-based practice have supported my learning	91
The remote teaching provided was effective to support my learning	97
I could get more help if I wanted it	91
The introductory session contained relevant information	82
The e-lectures presented the subject content adequately	97
The small group tutorials were useful in aiding my understanding of the subject	79

[a]vs. somewhat disagree/completely disagree

received with students finding it a useful application of their critical appraisal knowledge. The key area for improvement was more clarification of the difference between a systematic review and a critical appraisal, which was addressed for the 2021–2022 running of the workshop.

For the second year 2 workshop, the students particularly liked the group work, learning about clinical guidelines, and using their creativity to design a patient information leaflet. Some students would prefer the group presentations to be in person rather than online, to facilitate more engagement.

Students generally felt that the content of the third year 2 workshop was appropriate and provided a useful introduction to qualitative research methods, especially in relation to ethical aspects of this type of research. However, the students requested more clarity regarding the purpose and timetabling of the tutorial, which will be addressed for the next running of the workshop.

Formal feedback was not collected for year 4 EBP in 2020–2021. However, several improvements had previously been made based on student feedback, such as each project being assigned a mentor for guidance, and the provision of a report template.

Peer Feedback

For QRM in BDS18, all e-lectures were peer reviewed and updated based on feedback, plus there was a continuous cycle of peer review for small-group tutorials. It has not yet been possible to formally peer review the year 1 EBP course in BDS21 due to challenges imposed by COVID-19. However, the course is almost identical to the BDS18 QRM course, and also informal feedback is given by tutors after each tutorial; no substantial changes have been requested by tutors.

The Blackboard sites for all 3 year 2 EBP workshops were peer reviewed last year, minor suggestions for improvement were made, and these were incorporated into the workshops for this academic year; examples include the provision of clearer instructions for students and consistency of wording throughout the workshops. For year 4 EBP, informal feedback from mentors/markers is collected annually, and improvements made where needed, e.g. to the marking criteria.

In addition, all courses within BDS21 are evaluated during the Annual BDS Programme Review; no areas for improvement were identified for EBP.

External Examiner Feedback

BDS21 has a clinical and non-clinical external examiner for each year, and so far they have only reviewed assessments. No comments have been made relating to EBP.

4   Conclusions

Bristol Dental School's new undergraduate curriculum embeds statistics teaching within the theme of EBP. A range of statistical concepts are introduced and revisited throughout year 1 to year 4 of the dental degree, with published oral health research studies used to aid understanding and demonstrate the relevance to dentistry. Although this theme is led by non-clinical staff, clinicians contribute where possible.

The theme is delivered via a mix of online (mainly asynchronous) and face to face teaching methods, which use a variety of interactive activities. In addition to the application of knowledge and critical thinking, the main skills gained by the students during this theme are systematic searching of published literature, oral presentation of findings, writing for a lay audience, academic writing, teamwork, and peer assessment. Students generally engage very well with all the components of the theme, and the majority should be capable of applying EBP, ideally with some continued professional development, when they graduate as dentists.

Minor refinements of components within the EBP theme are expected based on future student and peer feedback. New GDC preparing for practice guidelines were due to be published but have been delayed due to COVID-19 [14]. Future modifications may therefore be necessary to ensure that all the relevant intended learning outcomes are met by the theme.

A similar model for teaching EBP could be implemented in other dental schools. Setting up the model is extremely time consuming, but once all the online resources and pre-prepared tutorial materials are established, future workload is minimised, and the course is sustainable even during pandemics. A further advantage is that in addition to the undergraduate students, postgraduate students and staff are also able to use any relevant resources. The main requirements for setting up and running a similar EBP model are a senior statistician who can devote a substantial amount of time to producing the resources, enough statisticians or epidemiologists to run tutorials, support from an e-learning specialist (e.g. for adding quizzes to recorded lectures, and setting up/maintaining a virtual learning environment such as Blackboard), and adequate time available in the curriculum.

References

1. Sackett DL, Rosenberg WMC, Gray JAM, Haynes RB, Richardson S. Evidence based medicine: what it is and what it isn't: it's about integrating individual clinical expertise and the best external evidence. Br Med J. 1996;312:71–2.
2. Evidence-Based Medicine Working Group. Evidence-based medicine: a new approach to teaching the practice of medicine. J Am Med Assoc. 1992;268:2420–5.
3. Richards D, Lawrence A. Evidence-based dentistry. Br J Dent. 1995;179:270–3.
4. Sellars S. How evidence-based is dentistry anyway? From evidence-based dentistry to evidence-based practice. Br Dent J. 2020;229:12–4.
5. General Dental Council. Preparing for practice and standards for dental education. 2011, revised version 2015. https://www.gdc-uk.org/docs/default-source/quality-assurance/preparing-for-practice-(revised-2015).pdf. Accessed 15 Nov 2021.
6. Overview of teaching of statistics within medicine and allied health sciences across UK universities. https://www.ed.ac.uk/usher/annual-meeting-teachers-of-medical-statistics-2018/overview-of-teaching-of-statistics-within-medicine. Accessed 15 Nov 2021.
7. Leary S, Ness A. Teaching research methods to undergraduate dental students. J Univ Teach Learn Pract. 2021;18(2):7.
8. Onwuegbuzie AJ, Wilson VA. Statistics anxiety: nature, etiology, antecedents, effects, and treatments – a comprehensive review of the literature. Teach High Educ. 2003;2:195–209.
9. Yilmaz MR. The challenge of teaching statistics to non-specialists. J Stat Educ. 1996;4:1.
10. Gill P, Baillie J. Interviews and focus groups in qualitative research: an update for the digital age. Br Dent J. 2018;225:668–72.
11. Flipped learning, Advanced HE Knowledge Hub. https://www.advance-he.ac.uk/knowledge-hub/flipped-learning. Accessed 22 Nov 2021.
12. Slot DE, Dorfer CE, Van der Weijden GA. The efficacy of interdental brushes on plaque and parameters of periodontal inflammation: a systematic review. Int J Dent Hyg. 2008;6(4):253–64.
13. CASP Tools. https://casp-uk.net/casp-tools-checklists/. Accessed 23 Nov 2021.
14. General Dental Council Learning outcomes review. https://www.gdc-uk.org/education-cpd/quality-assurance/learning-outcomes-review-process. Accessed 23 Nov 2021.

Teaching Medical Statistics Within the Context of Evidence Based Medicine

Matthew J. Grainge

1   History of Evidence Based Medicine

Much of the pioneering work in the field of Evidence Based Medicine (EBM) was conducted by David Sackett and colleagues [1, 2], although Sackett himself did not introduce the term (it was first coined by Gordon Guyatt in 1991). The commonly understood ethos of EBM was a change in paradigm away from unsystematic observations from clinical experience, towards a more focussed approach including efficient searching of literature and formal rules for its evaluation [3]. One of the first tangible outputs from the EBM initiative was the Cochrane Collaboration. Archie Cochrane was considered another one of the founders of EBM and early advocate of the use of randomised trials and systematic reviews to collate evidence. The latter was achieved after his death, when in 1993 a group of colleagues led by Ian Chalmers founded the collaboration in his name. The focus was originally on preparing systematic reviews of randomised controlled trials (RCT) and the first edition of the Cochrane systematic review of interventions was published in April 1995 [4].

    The National Institute for Health and Care Excellence (NICE) was formed in the United Kingdom (UK) shortly after this time. The purpose of the organisation is to provide guidance to improve health and social care. More specifically, it uses an evidence-based approach to guide healthcare commissioners for medical interventions, taking account of the cost of the intervention as well as evidence of effectiveness from RCTs and other sources. The concept of EBM has not been without criticism. Some of these criticisms include over-reliance on certain forms of evidence, in particular RCTs and meta-analyses, and that it does not meet its own empirical tests that clinical decision-making has been improved through EBM [5]. We can

M. J. Grainge (✉)
Lifespan and Population Health Academic Unit, School of Medicine, University of Nottingham, Nottingham, UK
e-mail: matthew.grainge@nottingham.ac.uk

D. J. J. Farnell, R. Medeiros Mirra (eds.), *Teaching Biostatistics in Medicine and Allied Health Sciences*, https://doi.org/10.1007/978-3-031-26010-0_3

also question the extent to which it impacts individual patients as their circumstances vary so widely. Furthermore, a lack of diversity in trial participants, which has been known to occur historically with RCTs, raises issues of external validity. Furthermore, the COVID-19 pandemic has highlighted further deficiencies in the EBM concept in that public health crises require the adaptation of sustained behavioural change without the luxury of being able to wait for evidence to emerge. The effects of such population-wide public health measures are difficult to establish through simple RCTs, still considered the gold standard for the application of EBM [6].

2 Purpose of Statistics Teaching

We must first consider the purpose of statistics teaching before considering it in the context of Evidence Based Medicine. The Outcomes for Graduates document states specifically that students should be able to "Interpret common statistical tests used in medical research publications" [7]. This would suggest that understanding of statistics would form part of a toolkit set up to enable students to critically appraise medical research evidence, which will inform decisions surrounding their future medical care. This could be the reason why compulsory modules containing statistics alone are less common in medical schools. However, there could be useful optional modules for any students who either have an interest in mathematics or statistics or who wish to pursue research in the future. There are broadly two reasons why we would teach statistics to medical undergraduates.

2.1 Doctors as Consumers of Research

This reason relates directly to their future lives as practicing clinicians. There is consensus from work over past two decades that medical students are more likely to be consumers rather than producers of research and that courses should focus on critical appraisal skills rather than the ability to analyse data [8, 9]. If a treatment is shown to work then the subsequent questions are "Will this treatment work well for my patients?" and "Is this research reliable or are there alternative explanations for any key findings?" These are sometimes termed external and internal validity respectively [10]. Critical appraisal is also quite a broad term and could mean anything from applying validated tools for assessing the risk of bias in a study [11], to identifying instances of potential research fraud (or undisclosed conflicts of interest). Strong numerical skills could also help our consumer of research to identify obvious errors in study reporting, e.g. results labelled the wrong way round or an effect estimate outside of its confidence interval. An equally important skill for future doctors is in the communication of risk to patients [12]. Understanding probability should be a key competency as is an appreciation that communicating relative measures of effect can cause unnecessary alarm if the absolute risk is low (and therefore the number needed to harm high in the case of an intervention).

2.2 Doctors as Producers of Research

This includes an immediate purpose of learning statistics for students undertaking a research dissertation as part of their degree. Whilst by their very nature research projects differ in the type and complexity of statistical methods they require, feedback from students enrolled on the Nottingham EBM module, described below, is along the lines that they are being taught statistics at the right time to do their research project. Other institutions also mention learning statistics at the time they embark on research dissertations.

3 EBM Teaching at Nottingham

At the University of Nottingham (UoN), medical students undertake a 6-month piece of research in the third year of their BMedSci degree, which contributes to 50% of their degree mark in year 3. The EBM module runs parallel with their dissertations to provide some wider context to their more specific piece of research, to explore how research gets put into practice, and to describe some of the specific methodologies which underpin this. The module is designed around Outcome 26 of the General Medical Council's (GMC) "Outcomes for Graduates" document covering Clinical Research and Scholarship [7]. The specified learning objectives of this module therefore coincide fully with the outcomes in the GMC document. The purpose of the module is to gear students towards being critical consumers of research, although there is little doubt that the module will also help them with their immediate role of being producers of research. The module was originally named "Research Methods" at inception in the mid-2000s, and was renamed Evidence Based Medicine when the author (MJG) took over as convener of the module in 2017. There has been discussion within our institution as to whether a series of didactic lectures are sufficient to impart all skills needed to succeed with their dissertations. Overall, it was felt this could not be achieved and that dissertation supervisors have an important role in providing tailored individual support to their students. As such, the name change was a reflection of the change in the ethos of the module to provide an insight into the wider context, which surrounds the specific piece of research that they are undertaking.

The topics covered in the module are listed in Table 1. All sessions are delivered by different teachers, all from the Schools of Medicine and Life Sciences at UoN. Teachers were chosen because they have specific expertise in the topic of focus, with a mixture of clinical and non-clinical academics. All sessions comprise a 50-min lecture (with two lectures for the Research Statistics topic) which take place in person. Each topic is delivered in a "standalone" way, meaning the material covered in one topic is not a pre-requisite for understanding other topics. This is important for logistical reasons. All lectures are delivered in an intensive 2-week interval at the start of the autumn semester. Therefore, this makes it impractical to deliver topics in a pre-defined order, due to timetabling issues and staff availability.

Table 1 Evidence Based Medicine module in year 3 of the BMedSci degree at the University of Nottingham: details from the module specification document (300 students annually)

Educational aim: The aim of this module is to introduce the concept of evidence-based medicine and review in depth many of the research methodologies, which underpin this. This will include ethical issues in animal and human research studies. Students build on previous learning and acquiring knowledge, skills and attitudes during the evidence-based medicine course that link to aspects of the outcomes for graduate specified by the GMC. They will learn about: the design of biomedical studies of various kinds; the collection, analysis and interpretation of data; and how to search for scientific information and how to critique biomedical studies and research papers. They will develop an understanding of the ethical considerations in medical research and the principles of academic integrity. Students will attend lectures and will be required to engage in private study, including the reading of scientific research papers.

Learning outcomes: The student will be able to meet outcomes described in the following paragraphs from the GMC Outcomes for Graduates (2018) document*: 5g, 19d, 25b, 25c, 25e, 25g, 25h, 26a, 26b, 26c, 26d, 26e, 26f, 26g, 26h, 26i, 26j.*

Topics covered (each covered in 50-min lecture):
  Research statistics (two lectures)
  Systematic reviews
  Bias in clinical trials
  Critical evaluation of the literature
  Qualitative research
  Principles of laboratory research
  Research ethics—Human studies
  Research ethics—Animal studies

Hence, the order of lectures often varies from year to year. As a 10-credit module, it is expected that all aspects of the module will consume 100 hours of study. We considered the idea of converting this to an online module several years ago, but feedback from students was that they welcomed the chance to attend lectures in-person at the start of academic year. This is especially the case, considering that students get little whole-group teaching in year 3, so this represents a good opportunity to catch up with classmates. During the COVID-19 affected years (2020–2021 and 2021–2022), teachers were given the option of delivering their lecture live via MS Teams or providing a recording of their lecture, which was followed by a live Q&A session in Teams.

The assessment for the module takes the form of an online 1.5-hour exam of multiple-choice type questions. This was open book in the two COVID-19 years when the exam was conducted remotely. The allocation of marks is assigned evenly across the topics listed above. We employ the modified Angoff method for standard setting of the assessment [13]. Multiple experts rate each question on the exam paper. For each question, the experts are asked to estimate the percentage of borderline (minimally competent) candidates they expect to answer the question correctly. Where there is initial disagreement, discussion takes place until a consensus is agreed. From these scores, a pass mark is estimated and all other marks in the cohort are scaled according to this. Since we adopted this approach, approximately 20% of students obtained a distinction (70% or higher) whilst a small number of students (<2%) fail (mark of less than 40%). Feedback is provided to students via a traffic light system where performance on questions related to each session objective is rated as red, amber or green. Progress throughout the module is monitored through participation in an online formative test containing questions of similar style and difficulty to the final assessment.

4 Statistics Teaching Within the EBM Module

Specific statistics teaching in the module builds on earlier teaching in year 1 of the BMedSci degree. Here students are introduced concepts of study design, sampling, hypothesis testing and communicating risk. Students receive 8 hours of teaching in year 1 on these topics in total, with statistical concepts entwined with these other concepts rather than as separate lectures. In the EBM module, students receive two 50-minute lectures on research statistics. The learning objectives and topics covered are listed in Table 2. We separate teaching into an overview of common statistical methods and measures of association and effect. Simple formulae such as odds ratios are provided and students are asked to apply a simple linear regression equation. However, the emphasis is on interpretation of results of statistical tests. This includes understanding of key assumptions underpinning particular statistical methods such as handling small cell frequencies when using chi-square. Students are also provided with a dataset to analyse using SPSS as part of their private study

Table 2 Content of two 50-min research statistics lectures that students attend as part of the Evidence Based Medicine module in year 3 of the BMedSci degree

Specific session objectives:
- Distinguish between continuous and categorical data
- Understand the concept of samples and how one makes inference back to the individual
- Understand what is meant by a confidence interval and the relationship between confidence intervals and hypothesis tests
- Describe and calculate measures of effect (odds ratios and differences between means)
- Understand how to present data descriptively (graphing, tabulations and summary statistics)
- Understand significance tests for comparing two groups of observations (t-tests and chi-square tests)
- Understand the assumptions involved with statistical tests for comparing two groups of observations
- Learn the difference between paired and unpaired data
- Understand when to use more advanced statistical methods (correlation, linear regression, ANOVA and logistic regression)
- Learn the difference between type 1 and type 2 error
- Appreciate how to avoid type 1 error
- Work out the size of a study

Statistical methods covered:
Descriptive analyses
Chi-square
t-tests
Correlation and regression
Logistic regression (concept only)
Sample size (concept only)

time, which they understand takes place outside of timetabled sessions. This enables then to work through chi-square, t-tests and regression which is done to help students develop a deeper understanding of these methods. The ability to carry out data analysis is not a key curriculum objective for most medical schools and against guidance from the GMC; we emphasise that this is the reason that they are asked to do this rather than us encouraging them to be data analysts [14].

As with all topics taught on the EBM module, there is an emphasis placed on self-directed learning. Therefore, we signpost students to openly accessible learning resources, which allow students to obtain a further deeper understanding of concepts which are difficult to fully grasp from didactic delivery. Examples of these include an illustration of how confidence intervals are constructed (https://rpsychologist.com/d3/ci/) and re-usable learning objects on various relevant topics developed by our institution (https://www.nottingham.ac.uk/helmopen). This learning also takes place in student's private study time. The combination of whole-group lectures combined with independent learning represents our preferred approach for teaching large groups of medical students. This is due to practical challenges in arranging small-group teaching sessions when timetabling a module, which takes place intensively over 2 weeks. In the two COVID-19 affected academic years, students were also provided with the opportunity to post any questions they had on the content of the lectures on a padlet board. These included questions such as "You said we should avoid multiple significance tests but how is this different from analyses such as multiple linear and multiple logistic regression?" which illustrate the depth of thought that some students were employing to comprehend the concepts taught.

5　　How Statistics Is Taught Alongside Other Disciplines

To explore the context in which statistics is taught in other institutions, I reviewed material from the "blue book" which is aligned with the annual meeting for teachers of medical statistics and contains publicly viewable information on how statistics is taught within each UK medical school (information as of September 2021) [15] (This documents on this website [15] were also reviewed in Chap. 1 and the interested reader is also referred to this chapter for more information.). Of 16 medical schools, which have an up-to-date blue book entry and where this entry is focussed on medical undergraduates, there is only one university which explicitly states there is no statistics teaching at the undergraduate level. Five universities only teach statistics in the context of standalone modules or as a statistics lecture series (this includes module titles such as "data science" which could be considered synonymous with statistics). Eight universities only teach statistics alongside other disciplines, including epidemiology, study design, literature searching, ethics and critical appraisal. The other two teach statistics both in a "standalone" way and as part of broader modules at different times during the course of the medicine degree. Of the 10 universities where statistics is taught alongside other disciplines, for 9 it is explicitly stated that critical appraisal and/or interpreting statistics in published research papers forms part of the module. This is in line with the above "Outcomes for Graduates" objective relating to how statistics should be taught. The term "Clinical Epidemiology" was used in two blue book entries and "Evidence Based Medicine" in four entries. However, the specific topics taught alongside statistics (listed above) tended to be covered frequently regardless of what title was applied to the module (e.g. "Research methods" or "EBM"). It should be emphasised that the above perspective is based solely on information obtained from the blue book. Just because statistics is taught in its own module or as part of an epidemiology module this does not mean that evidence-based medicine perspectives were ignored.

When statistics is taught alongside other topics, this will inevitably have a bearing as to how their statistical abilities and understanding will be assessed. A mixture of approaches was used for the ten blue book entries where statistics is taught alongside other disciplines highlighted above. These include multiple-choice questions (MCQs), written coursework, short answer questions, single best answer questions and application of formulae (for Poisson confidence intervals). This heterogeneity in approaches could reflect differences in class size, curriculum requirements, student background or the context in which the statistics is taught.

MCQs were the most commonly reported method of assessment (used in 5 of the 10 modules), which is likely to reflect the large group teaching scenario through which UG medical education takes place. This would suggest this assessment method is attractive in part for practical reasons. Newcombe writing in 1990 highlighted that MCQs are primarily used for retrieval of information stored from cerebral RAM and which does not correspond to the type of learning we wish to assess in our discipline [16]. However, it could be argued that statistics questions lend themselves particularly well to multiple-choice type (MCQ) questions if designed in such a way to require applied reasoning such as presenting data from a paper and asking students to interpret this. Overall, the description of assessment methods in the blue book entries was brief and did not address the issue of whether methods of assessment were compatible for both statistics and non-statistics components taught together in the module. This could indicate that an important area for future research is to provide a more empirical evidence base as to the extent to which statistical knowledge should be assessed when taught alongside other disciplines and whether failure to understand statistical concepts should ever be a reason for a student not being allowed to progress through medical school.

6   How Significant Is Our Teaching?

In my EBM statistics class, I used data from the dexamethasone study in reducing the risk of mortality in people admitted to hospital with COVID-19, chosen because this research was fresh in the public consciousness during the first lockdown in 2020 when I was preparing for that year's class [17]. As the purpose of this example was to illustrate a simple exposure vs. outcome relationship (i.e. chi-square) rather than the more sophisticated analyses presented in the paper, I provided the raw numbers from the paper in the form of a 2×2 table (Table 3).

Table 3 30-day mortality in patients admitted to hospital with COVID-19

Outcome exposure	Died	Alive	Total
Dexamethasone	482 (22.9%)	1622 (77.1%)	2104
Placebo	1110 (25.7%)	3211 (74.3%)	4321
Total	1592	4833	6425

Data from reference [17]
Raw data from the Recovery trial report (Dexamethasone and mortality following Covid-19)

Often when reviewing student dissertations (or even published research papers), these data will be frequently summarised along the lines of "There is a significant association between dexamethasone and mortality ($\chi^2 = 5.87$, $P = 0.02$)". We teach that the emphasis instead should be on the direction of the association (which group has the highest risk?) and the magnitude of the association (how much higher is it?), with statistical significance providing just a small amount of additional added value. So the interpretation should be "Those allocated dexamethasone had an estimated 14% reduction in risk of mortality within 30-days, with it being unlikely that this risk reduction is larger than 25% or that it increases the risk of mortality (odds ratio = 0.86; 95% CI 0.76–0.97)".

There has recently been heightened awareness on the extent to which we should be concentrating on null hypothesis significance testing (NHST). This is the result of guidelines issued by the American Statistical Association and a communication in the journal *Nature* with over 800 signatories, both of which are highly critical of the approach of NHST [18, 19]. Aversion to p-values and overreliance on use of the terms "significant" and "statistically significant" is something which is not new [20]. However, perhaps influenced by these recent statements there are now journals such as *Journal of Basic and Applied Psychology* which bans p-values entirely. In teaching statistics, we must consider this wider landscape and even prepare for a possible future without p-values. Indeed, this returns to the context of our teaching, and doing this as part of EBM provides an opportunity to place more emphasis on communication of risks (from both medical interventions and naturally occurring exposures) rather than inculcation of statistical tests through which use of NHSTs is unavoidable.

7   Future Challenges and Opportunities

Medical education has entered an exciting new phase, with applications to UK medical schools up by one-fifth in 2021. As an academic body, we are responsible for meeting the challenges of catering for this increasing number of medical undergraduates and ensuring that their teaching enables them to meet the competencies set out in the Outcomes for Graduates document. I highlight two challenges below which reflect my own observations but there are of course many others.

1. Curriculum design and peer evaluation
   At the University of Nottingham a recent re-design of the curriculum in years 1 and 2 has seen the introduction of large modules each focussed on an area of

medical anatomy. Teachers of medical statistics and related disciplines may not play a role in setting the over-arching agenda in which they teach. Having less control over the context in which we teach statistics is not necessarily a problem for the reasons I have highlighted here. We can maximise the impact and relevance of our teaching in the way we link it with other material covered at the same time. At the same time we should acknowledge that teaching of anatomy forms the bedrock of early years teaching in a medical degree and that development of a critical mind which is central to our discipline would be a natural development on this core teaching for future years, hence why we teach EBM in year 3. Peer evaluation of teaching initiatives can be used to provide teachers with an "external eye" to check that their teaching integrates well within the wider context.

2. *Encouraging a student-centred approach to learning*

A barrier to true student-centred learning is often the existence of assessment models which encourage and reward learning of taught material. The approach for our module of signposting students to freely available interactive resources is used to compensate for an otherwise largely didactic delivery style. However, if our aim is to encourage students to take responsibility for their learning, then teaching of statistics and EBM may lend itself more naturally to problem-based learning (PBL), which also facilitates statistics teaching to be more integrated with other parts of the curriculum [21, 22]. It should be noted, however, that incorporating statistics and research methods training into a PBL curriculum may present challenges not envisaged by those developing the curriculums. A study carried out in Australia identified one such barrier being the idea that the problem being addressed has to take the form of a patient vignette, which provides a difficult starting point for illustrating a concept such as a 2-sample t-test [23]. This paper is nearly two decades old and recent developments in PBL curriculums may have gone some way to countering some of these problems. This is evidenced in a recent study of year 3 medical students from Serbia, which provided RCT evidence that a PBL group obtained higher scores on an assessment than those receiving a blended learning approach only [24].

A more general consideration, which a PBL approach can help us to consider, is recognising that no two students are alike and thus we should be adapting our teaching for increasingly diverse groups of students, creating an environment where all students achieve their potential. Finally, the longer-term impact of the COVID-19 pandemic should be considered. It could be argued that our job of highlighting the value of medical data and importance of communication of risk has been made so much easier because of the media attention on the pandemic. An alternative view is that we may need to rise to the challenge to confront the increase in misinformation, prevalent in some aspects of the media.

8 Conclusion

In this chapter, I reviewed several areas of discussion, some of which date back to the 1970s, including teaching statistics primarily as a tool for critical appraisal. However, the subsequent decades have seen a growth in the popularity of EBM, communication of risk and personalised medicine among other things. Presentation and interpretation of numerical information is fundamental to all of these areas, thus allowing us to more convincingly illustrate the importance of our discipline and increase its impact through effective teaching alongside other disciplines. In 2002, Richard Morris posed the question "Does EBM force a change in the content of our teaching?" in a paper on this topic [21]. I would answer this by saying it is not intuitive to suggest that EBM will fundamentally change any of the underlying concepts we teach but that it does allow us to market its importance more effectively. We still have a responsibility to educate principles, which underpin our discipline such as p-values, but in this new landscape, we should emphasise that p-values and the hypothesis tests that produce them should never be relied on solely to the detriment of adequately summarising and presenting data and in the appropriate communication of risk.

References

1. Sackett DL. Clinical epidemiology. Am J Epidemiol. 1969;89:125–8.
2. Sackett DL. How to read clinical journals: I. Why to read them and how to start reading them critically. Can Med Assoc J. 1981;124:555–8.
3. Evidence-Based Medicine Working Group. Evidence-based medicine. A new approach to teaching the practice of medicine. JAMA. 1992;268:2420–5.
4. MacLehose H, Hilton J. Changes to the Cochrane Library during the Cochrane Collaboration's first 20 years. Cochrane Database Syst Rev. 2013;23:ED000050. https://doi.org/10.1002/14651858.ed000050.
5. Cohen AM, Hersh WR. Criticisms of evidence-based medicine. Evid Based Cardiovasc Med. 2004;8:197–8.
6. Greenhalgh T. Will COVID-19 be evidence-based medicine's nemesis? PLoS Med. 2020;17:4–7.
7. General Medical Council. Outcomes for graduates. GMC Publications. 2018. https://www.gmc-uk.org/education/standards-guidance-and-curricula/standards-and-outcomes/outcomes-for-graduates/outcomes-for-graduates. Accessed 24 Jan 2023.
8. Freeman JV, Collier S, Staniforth D, Smith KJ. Innovations in curriculum design: a multidisciplinary approach to teaching statistics to undergraduate medical students. BMC Med Educ. 2008;8:1–8.
9. Astin J, Jenkins T, Moore L. Medical students' perspective on the teaching of medical statistics in the undergraduate medical curriculum. Stat Med. 2002;21:1003–6.
10. Jüni P, Altman DG, Egger M. Systematic reviews in health care: assessing the quality of controlled clinical trials. Br Med J. 2001;323:42–6.
11. Sterne JAC, et al. RoB 2: a revised tool for assessing risk of bias in randomised trials. BMJ. 2019;366:1–8.
12. Sedgwick P, Hall A. Teaching medical students and doctors how to communicate risk. Br Med J. 2003;327:694–5.

13. Wyse AE. Comparing cut scores from the Angoff method and two variations of the Hofstee and Beuk methods. Appl Meas Educ. 2020;33:159–73.
14. Miles S, Price GM, Swift L, Shepstone L, Leinster SJ. Statistics teaching in medical school: opinions of practising doctors. BMC Med Educ. 2010;10:75.
15. Macdougall M. Overview of teaching of statistics within Medicine and allied health sciences across UK universities. Usher Institute, University of Edinburgh. 2021. https://www.ed.ac.uk/usher/annual-meeting-teachers-of-medical-statistics-2018/overview-of-teaching-of-statistics-within-medicine. Accessed 18 Nov 2021.
16. Newcombe RG. Evaluation of statistics teaching given to medical undergraduates. Stat Med. 1990;9:1045–55; discussion 1057–62.
17. The RECOVERY Collaborative Group. Dexamethasone in hospitalized patients with Covid-19. N Engl J Med. 2021;384:693–704.
18. Amrhein V, Greenland S, Mcshane B. Retire statistical significance. Nature. 2019;567:305–7.
19. Wasserstein RL, Lazar NA. The ASA statement on p-values: context, process, and purpose. Am Stat. 2016;70:129–33.
20. Sterne JAC, Smith GD, Cox DR. Sifting the evidence—what's wrong with significance tests? BMJ. 2001;322:226.
21. Morris RW. Does EBM offer the best opportunity yet for teaching medical statistics? Stat Med. 2002;21:969–77.
22. Sedgwick PM. Medical students and statistics challenges in teaching, learning and assessment. ICOTS8 Invited Paper 8; 2010.
23. Bland JM. Teaching statistics to medical students using problem-based learning: the Australian experience. BMC Med Educ. 2004;4:1–5.
24. Bukumiric Z, et al. Effects of problem-based learning modules within blended learning courses in medical statistics - a randomized controlled pilot study. PLoS One. 2022;17:1–14.

Teaching Null Hypothesis Significance Testing (NHST) in the Health Sciences: The Significance of Significance

Philip M. Sedgwick

Abbreviations

ASA   American Statistical Association
BASP   Basic and applied social psychology
NHST   Null hypothesis significance testing

1   Introduction

Traditional null hypothesis significance testing (NHST), incorporating the statistical null and alternative hypotheses plus critical level of significance of 0.05 (5%) needs no introduction. It is an essential component of all undergraduate and postgraduate statistics curricula. Statistical significance as denoted by $P < 0.05$ has become the cornerstone of decision-making in healthcare research; it is the gold standard for establishing if clinical significance exists—that is, if exposure to a personal behaviour or lifestyle, aspect of the environment, inborn or inherited characteristic is a risk factor for the development of a disease or condition, or determining if a newly developed medical strategy, drug, device, surgical approach, or alternative way of using a known treatment is effective. However, the idea that clinical significance can be implied from statistical significance represents misunderstanding of NHST [1]. Consequently, there have been repeated calls for statistics reform as to how NHST is used and applied. These calls for reform have focussed on the abandonment of the strict dichotomy of $P < 0.05$ versus $P \geq 0.05$, with

P. M. Sedgwick (✉)
Institute for Medical and Biomedical Education, St. George's, University of London, London, UK
e-mail: p.sedgwick@sgul.ac.uk

D. J. J. Farnell, R. Medeiros Mirra (eds.), *Teaching Biostatistics in Medicine and Allied Health Sciences*, https://doi.org/10.1007/978-3-031-26010-0_4

greater focus on the size of the observed sample effects and their importance in the clinical setting. Whilst reform is taking place, it is relatively slow and without universal agreement. For reform to occur and be sensible there needs to be enhanced understanding of NHST and *P*-values. It is suggested that NHST is poorly understood because students are not taught the philosophical concepts and ideas that underpin statistical theory and statistical inference. For change to occur, then without doubt the best place to start is in the classroom.

2   Teaching Statistics to Non-specialists

There has been much debate about the teaching of statistics to non-specialists. The teaching of statistics to undergraduate students in the healthcare sciences has traditionally presented challenges [2]. Students may not have any interest in the subject or even expect to study it at university, whilst students have varying abilities. For some, studying quantitative methods or statistics may create anxiety which can be shaped by their previous educational experiences [3]. Despite the best efforts of teachers, statistics has traditionally been considered by students as inherently mathematical and irrelevant [4]. Moreover, traditional methods of teaching statistics have been viewed as ineffective. This is because they have typically failed to present the relevance of statistical methods to future practice [5]. To help overcome these challenges there has been a change in how statistics is taught, with a shift towards statistical thinking and critical appraisal, rather than theory, formulae, and calculations. Moreover, this shift has focussed on helping students appreciate the relevance of statistics to future practice rather than learn theoretical contexts that do not have an obvious application.

Statistics is often thought of as being like pure mathematics in that it is an exact science. However, statistics is anything but an exact science. The fundamental concepts of statistics are centred on probability and uncertainty, and the discipline tends towards a discursive one. It is suggested that students fail to appreciate the fundamental principles that underpin statistical theory and inference, and this then contributes to challenges when learning and applying statistics.

2.1   Teaching NHST: An Overview

With a shift towards teaching that is focused on statistical thinking and critical appraisal, it would not be surprising if the philosophical concepts that underpin statistical theory and inference were no longer perceived as relevant in the curriculum. This may have contributed to NHST being typically taught as a recipe, and almost in a mechanistic way. In particular, the presence of statistical significance ($P < 0.05$), or otherwise, has become the only point of interest. Subsequently, statistical significance may have been taught or viewed as indicating clinical importance, not least to give simplicity to teaching and help give students a real-world application to their learning. Interestingly, there is not universal agreement if $P < 0.05$ or $P \leq 0.05$ indicates statistical significance; the critical level of significance used by journals differs, as does that presented in textbooks. Nonetheless, the most universally adopted cut-off is $P < 0.05$ (5%).

Hak advocated NHST should not be taught on undergraduate courses, and in particular not on introductory ones [6]. It was proposed that NHST was too difficult to teach, and too complex for students to understand. Moreover, the process of statistical significance based on $P < 0.05$ versus $P \geq 0.05$ simply encouraged dichotomous thinking. Furthermore, it was suggested there was no need to teach NHST and P-values. The rationale was that they did not provide any additional information to that of an absolute effect size and its associated 95% confidence interval. Post and van Duijn also acknowledged students found interpretation of statistical inference based on NHST difficult [7]. Nonetheless, they advocated that it was still necessary for students to appreciate the philosophical statistical concepts that underlie the principles of NHST and P-values in order for them to make contextual inferences based on effect sizes and confidence intervals.

Haller and Kraus discussed how, following teaching, the typical undergraduate student was unable to outline the underlying principles of NHST and the interpretation of a P-value [8]. Two possible sources for this were highlighted—that is, teachers of statistics and statistical textbooks. A survey reported that approximately 80% of psychologists who taught statistics had problems interpreting P-values. Therefore, failure to appreciate NHST and P-values was a problem that students shared with their teachers. This would suggest that in the continuous cycle where the student becomes the teacher, it is important that students are taught correctly so as to break the continued misuse and misunderstanding of NHST and P-values.

3 Approaches to Teaching NHST and the *P*-Value

It is suggested the greatest obstacle in understanding NHST and the P-value is the lack of awareness of the underlying principles of statistics. Indeed, few statistics texts even consider them. These principles are conceptual, and based on probability and uncertainty. Probability is an unintuitive and difficult concept for many [9]. To appreciate NHST and the P-value, students need more teaching of theoretical concepts and not less—as has been the trend in undergraduate and postgraduate courses over the last 30 years. Approaches to teaching NHST and P-values that have been found to be useful are presented below.

3.1 Teaching the Underlying Principles of Statistics

Students frequently lack an appreciation of the rationale of research and how statistical inference based on NHST is used to support research. The process of research involves selecting a sample from the population. In statistics the population is considered infinite, which is different to its general everyday meaning where it is used in a geographical sense. The statistical population is defined by the study criteria. A sample is selected because it is not possible to include an entire population, geographically at least, because of time and expense constraints. The sample is then used to make inferences about the population. Often anecdotal evidence based on an exploratory study utilising a small sample will suggest that a treatment is effective, or a personal behaviour or lifestyle, aspect of the environment, inborn or inherited characteristic is a risk factor for the development of a disease or condition. Nonetheless, any associations need to be explored in a so-called confirmatory study. Such studies utilise larger samples to ensure that the participants are sufficiently representative of the population. The objective is to make inferences based on the sample that are demonstrative of what would happen in the population. The process of making inferences about the population is typically achieved through null hypothesis significance testing (NHST). The aim is to establish if the sample supports the null hypothesis of clinical equipoise in the population—that is, if the treatments in a clinical trial are equivalent in effectiveness, or if exposure to a potential risk factor does not affect the risk of a disease or condition developing compared to the risk factor being absent. If the sample does not support the null hypothesis, then it is rejected in favour of the alternative, and it is inferred that the evidence does not support the position of clinical equipoise.

Teaching needs to be in context to give it a real-world application. To illustrate the fundamental philosophies of statistics, a study published in the *BMJ* is used as an example here [10]. The study was a double-blind, placebo-controlled superiority trial by design, the aim of which was to establish the effects of fish oil supplementation administered during pregnancy on infant body growth and composition. A total of 736 expectant mothers were recruited between 22- and 26-weeks' gestation. The primary outcomes included mean waist circumference (cm) of the offspring at age 6 years. At age 6 years, the mean waist circumference for the intervention group (fish oil supplementation) was higher than for control (placebo) ($n = 341$, mean = 55.5 (SD 3.8) cm *versus* $n = 347$, mean = 54.8 (SD 3.7) cm; mean difference = 0.7 cm; 95% CI: 0.14–1.26 cm; $P = 0.015$). The process of null hypothesis significance testing ascertains if the collected data support the null hypothesis which states clinical equipoise exists—that is, in the population no difference exists between the treatment groups in mean waist circumference of infants at 6 years. If the collected data do not support the null hypothesis, the process of NHST advocates the null hypothesis is rejected in favour of the alternative which specifies a difference exists in the population between the treatment groups in mean waist circumference of infants at 6 years. The statistical null and alternative hypotheses are formally specified in Box 1.

Box 1

Null Hypothesis: In the population from where the sample was obtained, no difference exists between the intervention (fish oil supplementation) and control (placebo) groups in mean waist circumference of infants at age 6 years.

Alternative Hypothesis: In the population from where the sample was obtained, a difference does exist between the intervention (fish oil supplementation) and control (placebo) groups in mean waist circumference of infants at age 6 years. The hypothesis is two-sided (two tailed); the mean waist circumference of infants at age 6 years for the intervention group (fish oil supplementation) could be greater, or smaller in magnitude than for control (placebo).

The fundamental philosophies of statistics are centred on the theoretical concept of repeated sampling from an infinite population, carried out under identical conditions and with the same sample size. The population is defined by the study criteria—that is, expectant mothers between 22- and 26-weeks' gestation. It is assumed that in the population there is no difference between the intervention (fish oil supplementation) and control (placebo) groups in the primary outcome as described under the null hypothesis of clinical equipoise (see Box 1). For each of the samples selected at random from the population, the difference between the intervention (fish oil supplementation) and control (placebo) groups in the mean waist circumference of infants at age 6 years is an estimate of the population parameter. Each sample will have a different sample estimate. Consider representing the theoretical concept of repeated sampling from an infinite population by arrows on a target board (see Fig. 1). The centre of the target board, the bullseye, represents the population parameter. Some arrows will be close to the bullseye whilst others will be towards the edge of the target board, reflecting sampling at random from the population. Obviously, only one study has been undertaken and the aim is for the arrow representing its sample estimate to be as close to the bullseye as possible, thereby minimising sampling error. Unfortunately, it is not known where the arrow for the single study has landed on the target board in relation to the population parameter. This idea of the target board serves to illustrate the concepts that underlie statistics, whilst helping to develop awareness of uncertainty that underpins statistical inference. In particular, there is nothing magical or special about the sample estimate obtained from the single study other than that it is the best (and typically only) estimate of the population parameter.

Fig. 1 Schematic representation of the fundamental philosophies of statistics, based on the concept of repeated sampling from an infinite population, taken under identical conditions, and with equal sample size. The centre of the target board, the bullseye, represents the population parameter, whilst the arrows represent the sample estimates obtained for the theoretical infinite number of samples. (Illustration by Damian Farnell, Cardiff University)

3.2   Sampling Distributions

Once the idea of infinite sampling has been introduced, students are able to appreciate the concept of sampling distributions. In the above example, the independent samples t-test could be used to undertake the statistical test of the null hypothesis. A different t-test statistic would be derived for each of the theoretical infinite number of samples. A histogram of these statistics would represent the t-distribution—that is, the so-called sampling distribution (see Fig. 2). It is important students are aware that the distribution is theoretical. Furthermore, the distribution is symmetrical about zero; the sample mean for the intervention group (fish oil supplementation) could be greater, or smaller in magnitude than that for control (placebo). The extremes of the distribution—the so-called tails—extend to minus and positive infinity. A test statistic of zero represents a sample with no difference between the treatment groups in the primary outcome. The greater the difference between treatment groups in the primary outcome—that is the sample mean of waist circumference for infants at age 6 years, the larger the magnitude of the absolute value of the test statistic. Subsequently, the test statistic will be further away from zero, indicative that the study provides less evidence to support the null hypothesis.

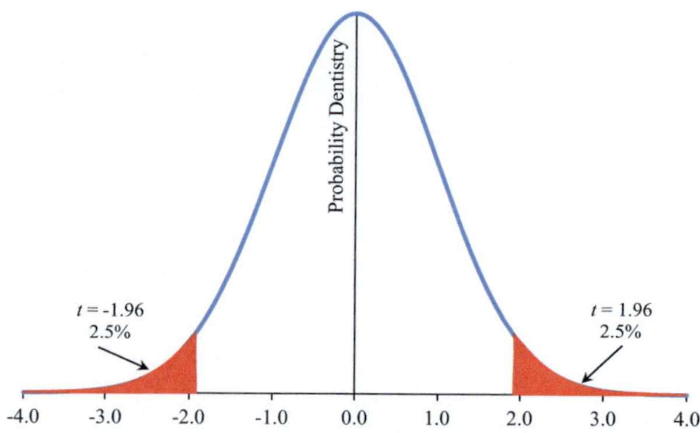

Fig. 2 For each of the repeated samples from the infinite population, a t-test statistic would be derived following comparison of the intervention group (fish oil supplementation) with control (placebo). The resulting distribution of the t-test statistics—the so-called t-distribution—is centred around zero. To obtain the P-value for the sample estimate observed in the single study, the proportion of the total area bounded by the histogram to the right of the absolute value of the test statistic is derived. This proportion is then doubled to obtain the resulting P-value; the proportion is doubled because under the alternative hypothesis the sample mean for the intervention group (fish oil supplementation) could be smaller, or larger than for control (placebo). Each of the red areas represents 2.5% of the total area under the curve; when combined they represent the so-called critical region of 5%. As sample size approaches infinity, the shape of the t-distribution approaches that of the normal distribution. In this situation, the test statistics of $t = -1.96$ and $t = +1.96$ define the critical region with 2.5% of the total area bounded by the curve in each tail. (Illustration by Damian Farnell, Cardiff University)

3.3   Deriving the *P*-Value

The *t*-test statistic for the statistical test comparing the intervention group (fish oil supplementation) with control (placebo) in mean waist circumference of infants at age 6 years for the above study was $t = 2.45$, with a resulting *P*-value of 0.015. The *P*-value was obtained by referencing the absolute value of the derived *t*-test statistic against the sampling distribution described above. The proportion of the total area bounded by the curve to the right of the derived *t*-test statistic ($t = 2.45$) equals 0.0075, i.e. 0.75%. Since the distribution is symmetrical, the area bounded by the curve to the right of the absolute value of the test statistic ($t = 2.45$) in the right-hand tail is equivalent to that bounded by $t = -2.45$ in the left-hand tail. The alternative hypothesis is two-sided, and therefore the *P*-value for the statistical test of the null hypothesis was derived as the sum of these two areas. With 0.0075 (0.75%) of the total area bounded by the curve in each of the tails, the resulting *P*-value is 0.015 (1.5%). Statistical significance is achieved for a study if the resulting test statistic is within the extremes of the tails of the distribution—that is, those 5% of samples with the largest differences between treatment groups (i.e. as represented by the 2½% of samples in the extremes of the tails). In Fig. 2, these two regions are shown as the red areas in the tails of the sampling distribution, collectively known as the critical region.

The *t*-distribution is symmetrical around zero, and bell-shaped like the normal distribution. However, the *t*-distribution has a flatter peak and heavier (i.e. thicker) tails. It was developed to account for the greater uncertainty associated with small sample sizes when estimating the population parameter. The shape of the *t*-distribution is dependent on the so-called degrees of freedom, which are derived as the total number of participants in the two treatment groups minus the number of groups being investigated: for the example above the degrees of freedom equal 736 minus 2, i.e. 734. As sample size approaches infinity, the shape of the *t*-distribution approaches that of the normal distribution. In such a situation, the test statistics of -1.96 and $+1.96$ define the critical region with 2.5% of the total area bounded by the curve in each tail. In smaller samples, the absolute value of the *t*-test statistic that defines the critical region is larger in magnitude. For example, when the sample size is 60 the *t*-test statistics that define the critical region are the values of -2.00 and +2.00.

3.4   What Does *P* < 0.05 Mean?

By illustrating the underlying principles of statistics as described above, it helps students appreciate how a *P*-value is obtained. The derivation of the *P*-value is based on a theoretical concept within the population—that is, under the null hypothesis, which proposes that in the population there is no difference between the treatment groups in the primary outcome. Statistical significance ($P < 0.05$) is not magical—it is simply achieved if the study observed is one of the extreme 5% of the theoretical samples following repeated sampling. Statistical significance is therefore a mathematical concept. However, teaching and textbooks typically describe the process of NHST as a recipe, utilising a black-box approach with no instruction as to how the P-value is derived. Moreover, it is implied, often incorrectly, that if statistical significance is achieved ($P < 0.05$) then it denotes clinical importance. The leap in inference from statistical significance to clinical importance is typically seen by students as magical, which may be because of how the topic is usually taught.

Students should be encouraged to consider the clinical significance of the results of a study, regardless of whether statistical significance ($P < 0.05$) is achieved. It is important to inspect for potential clinical importance in the data collected based on the sample estimates of the population parameters. For the above study, there was a statistically significant difference between the intervention group (fish oil supplementation) and control (placebo) in mean waist circumference of infants aged 6 years. However, it is not easy to appreciate if the difference between treatment groups in mean waist circumference—an increased mean difference of 0.7 cm in favour of the intervention—is important clinically. The latter is a difficult question for students to consider in their teaching, not least because they are unlikely to be experts in the speciality of early childhood development. Students generally find it difficult to consider the application and importance of statistical inferences from a teaching scenario in the clinical or healthcare setting [11].

3.5   History of NHST

To appreciate the current debate around inferences centred on the bright line rule of statistical significance ($P < 0.05$ versus $P \geq 0.05$) and to overcome these challenges in the future, it is imperative that students are made aware of the history of the process of NHST [12]. Students should therefore be given a short introduction to the development of NHST.

Traditional NHST as described above is a single process based on two discrete theories proposed by Fisher in 1925 [13], plus Neyman and Pearson in 1933 [14]. Fisher proposed the null hypothesis and P-value. The P-value was the strength of the evidence provided by the sample data to support the null hypothesis. Although Fisher advocated $P = 0.05$ (5%) for statistical significance, it was not as an unconditional cut-off. His intention was that statistical significance should be used as a tool to indicate if the study results warranted further investigation. In particular, interpretation of the study results was meant to be subjective and it was for the researcher(s) to decide upon the contextual importance of the collected data. Neyman and Pearson subsequently suggested the concept of hypothesis testing, advocating the null hypothesis could not exist without an alternative one. Furthermore, they suggested the probabilities of making incorrect decisions— namely, type I and II errors—should be set in advance. A type I error would occur if the statistical null hypothesis of clinical equipoise was rejected in favour of the alternative, when the position of equipoise existed in the population. A type II error would occur if the null hypothesis of clinical equipoise was not rejected in favour of the alternative, when the position of clinical equipoise did not hold in the population and a difference between groups in outcome existed. Such errors can occur because of selection bias. Neyman and Pearson never advocated an absolute cut-off of 5% (0.05) as the maximum probability of a type I error. It is not entirely clear how, why, or when the two theories were amalgamated to give traditional NHST. Fisher and Neyman-Pearson were fiercely opposed in their schools of thought and the combination of their theories, without doubt, represents a misunderstanding. Whilst the theories of Fisher and Neyman-Pearson are discrete, they have parallels which no doubt led to their combination. In particular, the critical region of the Neyman-Pearson theory—that is, the maximum probability of making a type I error—can be defined in terms of Fisher's P-value and proposed statistical significance ($P < 0.05$). Furthermore, traditional NHST has become a convenient way of declaring the presence of clinical significance even if it is an incorrect inference.

3.6   Developing a Culture of Uncertainty

When teaching statistical inference based on NHST and P-values, it is important to remind students that the underlying principles of statistics are founded on probability and uncertainty. Data is evidence that we can use to derive a P-value; as a probability it provides a measure of the strength of evidence that the sample estimate provides to support the null hypothesis. If statistical significance does not exist (i.e. $P \geq 0.05$), the inference is that there is no evidence to reject the null hypothesis in favour of the alternative. Furthermore, if statistical significance ($P < 0.05$) exists there is little evidence to support the null hypothesis and it is rejected in favour of the alternative. During this process there should be avoidance of indicating statistical hypotheses are true or false, not least because they can never be proven or disproven. Statistical inference is based on the concept of repeated sampling from an infinite population, under identical conditions and with the same sample size; the samples will give different sample estimates for the population parameter. In practice only one study is undertaken, and the ensuing inference based on statistical hypothesis testing could be a type I or type II error.

Typically, it is challenging for students to appreciate the concept that neither the null hypothesis nor the alternative is ever accepted as part of the NHST process. As described above, it is imperative that students are made aware of the inferences that can be made from NHST. Altman and Bland coined the expression *"Absence of evidence is not evidence of absence"* [15]. That is, whilst statistical significance may not exist, the inference is that there was no evidence of an effect—not that one does not exist in the population [16]. The idea of evidence is not absolute in statistical inference, which is different to the way it is perceived in a court of law.

It is not uncommon for researchers to use the words "positive" and "negative" when referring to statistically significant and non-statistically significant results. This phraseology should not be used because the words "positive" and "negative" have specific meanings in everyday language based on good and bad outcomes, or more generally success and failure. The results of research should never be considered as such, and as teachers we must be careful of our language as it may encourage students to misinterpret the objectives of NHST, intentionally or otherwise.

Students find it challenging to recognise that there is a difference between statistical inference and the "truth". Whether healthcare professionals can live with the uncertainty in clinical effectiveness based on probability is not clear, not least because it is not what they or their patients wish to know. Healthcare is underpinned by the need to know, for example, whether a treatment or therapeutic regimen is superior to standard care or placebo. The dichotomy offered by traditional NHST is a convenient one in this thinking. For most clinicians, "…*a statistically significant P-value is the end of the search for truth*" [17]. However, NHST cannot provide this information.

4   Statistics Reform

The misuse and misinterpretation of NHST—in particular, the misunderstanding that clinical importance can be inferred from statistical significance, has been debated by the scientific community for decades. In 1951, Yates proposed the eventual aim of researchers was to determine if statistical significance existed, and they paid little attention, if any, to the sample estimates of the population parameters being investigated [18]. This debate has intensified within the last decade. In 2015, the editors of the journal *Basic and Applied Social Psychology* (*BASP*) banned statistical significance based on the bright line rule of $P < 0.05$ versus $P \geq 0.05$ [19]. In addition to statistical significance, the ban included P-values, test statistics, and proclamations about significant differences or lack thereof.

In 2016, the American Statistical Association (ASA) issued a statement providing guidance on the context, process, and purpose of P-Values [20]. The ASA's statement was considered a more balanced approach to statistics reform than that of the editors of *BASP* a year earlier [19]. The emphasis was supporting the informed use of statistical inference based on NHST, rather than banning it simply because it is misused. Several years later, *The American Statistician* published a special issue entitled "*Statistical Inference in the 21st Century: A World Beyond p < 0.05*" [21]. The objective was to provide continuing guidance on the use of NHST, P-values, and statistical inference. The main message was that the concept of "statistically significant"—that is, $P < 0.05$—should no longer be used. Whilst the journal *The American Statistician* is the official publication of the ASA, it is important to recognise this message represented the opinions of the editors of the special issue and not those of the ASA. In the same year, *Nature* published a prominent article in which the authors proposed that statistical significance based on the categorisation of $P < 0.05$ versus $P \geq 0.05$ should be abandoned [22]. It was claimed that such categorisation has led to "*…hyped claims and the dismissal of possibly crucial effects*". However, it was not proposed that P-values be banned. Instead, P-values should be used in addition to an emphasis upon the sample estimates and their precision in respect of their importance in healthcare.

As the debate regarding the misuse and misinterpretation of NHST has been ongoing for decades, it is not clear why the application of statistical significance ($P < 0.05$ versus $P \geq 0.05$) has continued in decision-making. George Cobb highlighted why progress may have been limited, when at an ASA forum in 2014 (personal communication) he commented *"We teach it because it's what we do; we do it because it's what we teach"* [23]. In particular, Cobb emphasised the circularity in the application of NHST. Therefore, the best place to start to have an informed debate about statistics reform regarding NHST is in the classroom.

5   Concluding Comments

It is suggested that teaching and learning statistics is difficult because the underlying principles of statistics are conceptual and based on uncertainty. Such principles are typically omitted from teaching and textbooks. Statistical inference is based on probability, which is a difficult concept for many. The interpretation of the P-value is not simple and intuitive. These challenges may have contributed to NHST being habitually taught mechanistically such that the ultimate point of interest in this recipe has become the presence of statistical significance, or lack of it. To add to the difficulties of teaching and learning about NHST and P-values, students find it challenging to consider the direct application of statistical inferences from teaching scenarios as they are not necessarily experts in the clinical and healthcare specialties illustrated. For debate about statistics reform to be informed, teachers need to consider their approaches to learning and teaching.

References

1. Sedgwick PM, Hammer A, Kesmodel US, Pedersen LH. Current controversies: null hypothesis significance testing. Acta Obstet Gynecol Scand. 2022;101:624–7. https://doi.org/10.1111/aogs.14366.
2. Sedgwick PM. Medical students and statistics: challenges in teaching, learning and assessment. In: Reading C, editor. Data and context in statistics education: towards an evidence-based society, proceedings of the eighth international conference on teaching statistics (ICOTS8, July, 2010), Ljubljana, Slovenia. https://iase-web.org/documents/papers/icots8/ICOTS8_4E2_SEDGWICK.pdf.
3. Peiró-Signes A, Trull O, Segarra-Oña M, García-Díaz JC. Anxiety towards statistics and its relationship with students' attitudes and learning approach. Behav Sci. 2021;11(3):32.
4. Altman D, Bland JM. Improving doctors' understanding of statistics. J R Statist Soc A. 1991;154(2):223–67.
5. Mustafa RY. The challenge of teaching statistics to non-specialists. J Stat Educ. 1996;4:1.
6. Hak T. After statistics reform: should we still teach significance testing? In: Makar K, de Sousa B, Gould R, editors. Sustainability in statistics education. In: Proceedings of the ninth international conference on teaching statistics (ICOTS9, July, 2014), Flagstaff; 2014.
7. Post WJ, van Duijn MAJ. Teaching hypothesis testing: a necessary challenge. In: Makar K, de Sousa B, Gould R, editors. Sustainability in statistics education. Proceedings of the ninth international conference on teaching statistics (ICOTS9, July 2014). Flagstaff; 2014.
8. Haller H, Krauss S. Misinterpretations of significance: a problem students share with their teachers? Methods Psychol Res. 2002;7(1):1–20.
9. Spiegelhalter D. Why do people find probability unintuitive and difficult? 2022. https://nrich.maths.org/7326. Accessed 31 Jan 2022.

10. Vinding RK, Stokholm J, Sevelsted A, Sejersen T, Chawes BL, Bønnelykke K, Thorsen J, Howe LD, Krakauer M, Bisgaard H. Effect of fish oil supplementation in pregnancy on bone, lean, and fat mass at six years: randomised clinical trial. BMJ. 2018;362:k3312.
11. Aquilonius BC, Brenner ME. Students' reasoning about p-values. Stat Educ Res J. 2015;14(2):7–27.
12. Kennedy-Shaffer L. Before p < 0.05 to Beyond p < 0.05: using history to contextualize p-values and significance testing. Am Stat. 2019;73(Suppl 1):82–90.
13. Fisher RA. Statistical methods for research workers. 1st ed. Edinburgh: Oliver and Boyd; 1925. p. 1925.
14. Neyman J, Pearson E. On the problem of the most efficient tests of statistical hypotheses. Philos Trans Royal Soc. 1933;1933(231):289–337.
15. Altman DG, Bland JM. Absence of evidence is not evidence of absence. BMJ. 1995;311:485.
16. Sedgwick P. Understanding why "absence of evidence is not evidence of absence". BMJ. 2014;2014(349):g4751.
17. Banerjee A, Jadhav SL, Bhawalkar JS. Probability, clinical decision making and hypothesis testing. Ind Psychiatry J. 2009;18(1):64–9.
18. Yates F. The influence of "statistical methods for research workers" on the development of the science of statistics. J Am Stat Assoc. 1951;46:19–34.
19. Trafimow D, Marks M. Editorial. BASP. 2015;37(1):1–2.
20. Wasserstein RL, Lazar NA. The ASA's statement on p-values: context, process, and purpose. Am Stat. 2016;70(2):129–33.
21. Wasserstein RL, Schirm AL, Lazar NA. Statistical inference in the 21st century: a world beyond p < 0.05. Am Stat. 2019;73(Suppl 1):1–401.
22. Amrhein V, Greenland S, McShane B. Retire statistical significance. Nature. 2019;567:305.
23. Cobb G. Personal communication. ASA Discussion Forum, Mount Holyoke College; 2014.

Teaching Conceptual Understanding of p-Values and of Confidence Intervals, Whilst Steering Away from Common Misinterpretation

Hilary C. Watt

1 Introduction

1.1 Concern That Statistical Inference Is Hard to Understand

Statistics has a reputation as a hard subject. University students and researchers often have misunderstandings of some aspects of statistical inference [1–5]. Some professors of statistics acknowledged to me that they only gradually understood the principles underlying statistical inference as they started out on their careers. Some students would report that they did not see the practical relevance of statistical inference or understand its purpose via conventional teaching methods. For example, some students believe that 95% of observations lie within 95% CIs, which demonstrates a fundamental misconception over the purpose of CIs [2]. There may be confusion over why we use an interval to estimate odds ratios (and other statistics) rather than a single value. This all provides evidence for a need for greater focus on conceptual understanding of CIs and of p-values when teaching statistics.

The distribution of sampling statistics is known to be a challenging concept [6]. Tutors often feel obliged to teach several different sampling distributions to their students, to enable them to understand the calculations underpinning different statistical techniques. Statistics tutors and their students may feel obliged to spend substantial time and effort on this technical content. The language used to interpret p-values and CIs might not help all students to develop conceptual understanding, for reasons that are explained in the relevant sections below. Therefore, I call for attention being given to creating/choosing "definitions" and interpretations that foster conceptual understanding.

H. C. Watt (✉)
Department of Public Health and Primary Care, Imperial College London, London, UK
e-mail: h.watt@imperial.ac.uk

1.2   Excessive Focus on $p < 0.05$ and Declaring Results as "Significant"

Many courses teach students to make different interpretations according to whether $p < 0.05$ [7], reflecting common practice in research publications, and simplifying teaching and marking. Students may be taught to declare results as "significant" or not, which informally implies large enough to matter; it does not accurately reflect the information about statistical uncertainty encompassed in a p-value. The term "compatibility" is widely promoted as an alternative, with "surprise" and "statistical clarity" also suggested [8–11]. When interpretations using "significance" are emphasised in statistics courses, this may perpetuate bad practice in published research [7, 12]. Without the use of other teaching strategies, students may have little chance of developing understanding of p-values that would enable them to make more skilful interpretations. Many courses probably do include other strategies; to my knowledge, there are no surveys of what techniques are used. The highly respected ASA p-value statement [7] calls for a focus on scientific, rather than statistical inference that takes account of many different factors including the p-value, which is interpreted on a continuous scale [7, 13–15]. Full implementation of this approach may require more teaching time than is available. From personal experience, many courses do inform students that the $p < 0.05$ cut-off is arbitrary, and that p-values should be interpreted according to the size of p-value. However, students might still focus on the simplest $p < 0.05$-based rule because this is a complex subject that many find confusing. The fact that such rules may be appropriate for primary research outcomes, but not for speculative testing [12], is another issue that may not always be sufficiently emphasised in statistics teaching.

One justification to focus on a specific cut-off is that clinicians need to make decisions, for instance on whether to promote a drug treatment, based on results of randomised trials. However, the guidance on scientific inference still applies. For trials, the $p < 0.05$ cut-off is taken seriously but not rigidly, in practice, alongside broader considerations. A fixed p-value cut-off ("alpha", usually taken to be the traditional value of 0.05) is required for sample size calculations. I am not aware of any courses that avoid referring to the $p < 0.05$ cut-off completely. Students who read the research literature need to know that the term "significance" usually implies $p < 0.05$. I aim to avoid the term "significance" (except to explain how others use it), but I do use "$p < 0.05$" (as in this chapter).

1.3   Evidence of Over-generalising Conclusions from p-Values

There is evidence that many people make inappropriate inferences from p-values [7, 16, 17]. Many forget that we have certainty amongst study participants, with results based on direct calculations (calculated odds ratios and similar). One research study presented researchers with calculated results amongst two groups of study participants—many researchers erroneously reported that they did not know whether there was a group difference amongst participants, when presented with $p > 0.05$ for this comparison, but not when presented with $p = 0.01$ [14, 18]. They appear to have forgotten that p-values are not relevant to results amongst study participants, even when the wording of the question emphasised this focus. Around 80% of researchers published in the *New England Journal of Medicine* made this mistake, with around half of the authors of the *Journal of American Statistical Association* who are (primarily) statisticians also making this mistake. Whilst many students with statistical training made this error, students without statistical training were far less likely to make this error and were not influenced by p-values [14, 18].

The demonstrated beliefs that p-values relate to uncertainty amongst participants implies lack of knowledge of what p-values do represent. P-values reflect uncertainty in population estimates, when considering subjects to be selected at random from some broader population. Focusing on the population is a tool for accounting for imprecision, based on random choices depending on who is selected into the study. This evidences the need to use more skilful teaching strategy so that more students develop this knowledge.

1.4   Concern Over Standards of Statistical Interpretation in the Applied Literature

A major concern is that of ignoring associations amongst study participants, and describing them as "no association", when $p > 0.05$ for this comparison [7]. The opposite issue, of presuming that any results with $p < 0.05$ reflects some genuine association, is equally problematic. This leads to some people basing compatibility of research results, across different studies, primarily on whether each study individually reports $p < 0.05$ for the comparison of interest [16, 19]. Some people may declare results to be incompatible merely because only one study reports $p < 0.05$, even when both studies report very similar odds ratios and with greatly overlapping confidence intervals. Hence, this excessive focus on whether $p < 0.05$ can lead to a lack of ability to make sensible research conclusions across a body of research. This may have practical implications for medical treatment and lifestyle advice given to patients.

The American Statistical Association (ASA) has focused such concerns for the past several years. Their highly respected statement on p-values [7] provides advice on appropriate interpretation of p-values, based on expert consensus. They also promote a focus on scientific inference, taking account of many different sources of evidence, far broader than p-values and CIs.

1.5   Concerns That Odds Ratios (and Similar) and Are Ignored

For much of this chapter, I refer to odds ratios, since this is one of the most common measures of association reported in the literature. However, any other measures of association, such as relative risk, incidence rate ratios or differences in mean, could be used in its place.

Many people feel that measures, such as odds ratios and their CIs, deserve more attention than they currently receive [7, 16]. An excessive focus on whether $p < 0.05$ can only be misleading. The fact that many people who are statistically trained failed to recognise that we know results amongst study participants, based on calculations from data, is further evidence that such clarity is currently lacking [18].

When teaching statistics, many people refer to study participants as the "sample", which might make clinicians think of blood samples, or morsels of free food tempting them to buy more. Some refer to calculated odds ratios as "point estimates" or "best estimates". This terminology considers participants to be a random "sample" of some theoretical population, and the study participants' data to be "best estimates" of population parameters. Yet the population "that participants were selected at random from" rarely exists, but is merely theoretically defined to allow for calculation of p-values and CIs [20]. Referring to study participants directly as participants or as subjects can only help to make statistical interpretations clearer and potentially easier to understand.

1.6 Concerns That CIs Are Not Understood and/or Ignored

Given that many students find it challenging to understand inference [1, 2, 21, 22], it is helpful to review how CIs are interpreted. The formal definition of CIs refers to taking many repeated samples from some pre-determined population and constructing CIs for odds ratios (or similar) calculated from each sample [23]. On average, 95% of these random samples will produce CIs that include the population odds ratio.

Many people may find this formal CI definition unsatisfactory or counterintuitive. This is in part because we naturally want to find an interval where we can claim 95% probability that the population odds ratio lies within it. Yet the frequentist approach does not enable this, leaving us with CI logic that feels back-to-front, when interpreted using ideas derived from "probability". CIs tell us about compatibility of different possible values for population odds ratio with our data; use of "compatibility" may make such interpretations feel more natural [8]. Frequentist statisticians (unlike Bayesians) do not combine this compatibility information with information on which values for odds ratios are more probable based on prior evidence. Hence, "95% probability" cannot be claimed.

The CI definition may also feel unsatisfactory, since it speaks of taking many random samples from some population, yet taking repeated random samples is not a study design used in practical research. Furthermore, study participants are rarely selected at random from any population. In my view, it does not feel like a definition because it does not describe the purpose or intention behind CIs. It does describe a crucial property of CIs that 95% of estimates from these random samples include the population parameter being estimated.

There is widespread use of informal interpretations of CIs, which feel more intuitive and may be easier for students. Informally, statistics tutors may say: "we are 95% confident that population odds ratio lies between 0.95 and 2.43". This is a bit of a fudge, of course, as "95% probability that population odds ratio lies within…" is not generally acceptable. My experience is that students are comfortable with this form of words and can very easily use such formats, except that they may forget the word "population". There is evidence that many students do not understand what is meant by "population" [8, 24–26], so they would not then fully understand these informal CI interpretations that rely on an understanding of the "population".

2   Proposal

2.1   Setting Where Teaching Was Delivered

These teaching ideas were developed and delivered primarily on postgraduate introduction to Statistics course on the Master's in Public Health (MPH) at Imperial College London. Some students had little or no knowledge of statistics at entry, whereas others had knowledge of three or four regression methods. The course was 1 day per week over 10 weeks to around 70 students each year, divided into one (in 2019), two (in 2021) or three (in 2018) workshop groups, with one lead tutor and a few assistant tutors per group. 2020 was entirely online due to Covid-19, with software taught changing from Stata to R. Data from 2021, but not 2020, are reported on here. The graphs (presented below), designed to develop conceptual understanding of CIs and p-values, were introduced in 2018 (in Stata software workshops) and additionally into lectures in 2019. Transparent explanatory CI and p-value interpretations were incorporated into workshop summaries of core concepts (in 2018) and into lectures (in 2019) and throughout detailed solutions in 2018 and 2019. In 2018 and 2019, students had two lectures in person of around 60 min each, prior to workshops. In 2021, students were required to engage with online material including short lectures, reading material and exercises, including R code and exercises with some interpretation. Day one was introductory, with motivation for the importance of the subject, and focus on types of data and additionally (in 2018 and 2019) descriptive statistics (taught on day 2 in 2021). They were taught CIs on their second day each year, and p-values (with paired and unpaired t-tests, and chi-squared tests) on their third day. The further 7 days focused on linear, logistic and cox regression analyses, including interactions and in-depth checking of assumptions and model building and implications of missing data, with associated software code. CIs and p-values interpretations were taught, as applied to interpreting regression coefficients, throughout those weeks. They had statistical interpretation workshops of around 90 min each week, followed by Stata/R workshops of similar length, which included statistical interpretation. The CI and p-value transparent explanatory interpretations incorporated into the R workshop statistical interpretations in 2021 are presented below. These used slightly different forms of wording to those used in 2018 and 2019. This statistics module was one of four modules taught over their 10-week term.

2.2 Choose Language That Emphasises Calculated Results Amongst Study Participants

I choose to refer to odds ratios "in study participants". This straightforward clarity may encourage people to pay greater attention to these calculated results. This contrasts with the common practice of referring to participants via their relationship to the theoretical inference population, which implies calling them "estimated" odds ratios or odds ratios in the "sample". Whilst teaching research is required to explore this, our anecdotal experience suggests that combining this with suggested CI interpretations below makes CIs easier to understand. This language was used throughout software workshops solutions in 2018, 2019 and in 2021, albeit the form of CI "definition" chosen in 2018 and 2019 was a minor variation on those reported below, as previously published [15].

2.3 Confidence Intervals (CIs) Interpretations That Promote Understanding

The concept of the population is so fundamental and sufficiently challenging that it is worth describing over and again. As statisticians, we often teach students to acknowledge the assumptions underlying their analyses, as promoted by ASA p-value statement [7]. A key assumption which is rarely reported is that of random sampling of study participants from the population. The following was incorporated into R workshop solutions in 2021.

Odds ratio is 1.52 in study participants. We are 95% confident that population odds ratio lies between 0.95 and 2.43 (assuming random sampling of study participants from this greater population).

Notice the contrast between the results in study participants and referral to the population in the above CI interpretation. Some students reported that they did indeed appreciate the clarity of these interpretations, despite being told that they did not need to provide such interpretations in their exams. Whilst the above successfully describes the population, it does not describe the purpose of introducing this concept. An additional phrase (below) may be added to further describe what CIs represent; this is designed to provide a practical interpretation that is not dependent on the concept of the population (since this usually lacks a concrete context-dependent definition). Reference to "probability random chance" is to avoid misconceptions of the word "random", which is used in everyday speech to imply haphazard or out of the ordinary [27]. Students can be taught that "probability random" samples have known properties for long run averages and distribution shapes.

This interval reflects the precision of our estimate of the population odds ratio, with precision based on probability random chance alone.

I refer to the two italicised sections above combined as a transparent explanatory CI interpretation, with the intention that it clearly describes the CI concept. The above has the merit that it is readily compatible with teaching material that interprets CIs using the more standard informal phrase "we are 95% confident …" without the added assumption. This version was used for compatibility with pre-prepared online material, developed before these transparent explanatory interpretations were incorporated into teaching material.

The ASA p-value statement [7, 8] recommends the term "compatibility" for p-values. CIs can be defined as ranges of values for population parameters that result in $p > 0.05$ for comparison with study data (for 95% CIs, assuming technical details of calculation methods correspond appropriately). Because of this close link between p-values and CIs, the same terminology is relevant to both. Use of "compatibility" rather than "confidence" better reflects the evidence contained within a CI. I am curious about whether the more accurate reflection of the evidence in the following CI interpretation might help students to make sense of statistics, or whether its greater complexity might hinder their understanding.

> *Calculated odds ratio is 1.52 in study participants. Study data are readily compatible with random selection from a population with odds ratio between 0.95 and 2.43 (giving 95% coverage of compatible values, conditional on assumptions of the underlying calculations being true).*

Whilst such phrases were occasionally used in teaching in 2021, I lack evidence over how students relate to them. Students might, of course, have different preferences. The version that most aids student understanding might depend on the level of the course.

Because many students struggle to understand CIs, students may be asked to interpret many CIs throughout a statistics course. Most courses would report a standard CI definition or chosen form of CI interpretation repeatedly, including in workshop solutions, applied to different situations. This allows students to learn to answer "correctly". Therefore, there is value in choosing the wording used to interpret or "define" CIs wisely, so that their repetition supports the aim of improving student understanding. Before using this approach on our course, different short paragraphs were used to describe CIs in different parts of student solutions. This approach did not go down well. Students appeared to be overwhelmed by them and complained about their length. However, long solutions were perceived as very useful when written in a very systematic manner, including two chosen forms of transparent explanatory CI interpretation.

Better understanding of CIs may encourage more focus on them in the published literature.

2.4   Potential Misunderstanding When Focusing on the distribution of Sample Means

The focus on the following sections is on teaching CIs for means, chosen for its simplicity and visual appeal. In particular, data can be presented in histograms, with the CI for the mean overlaid onto them.

A diagram that is routinely shown to students is the distribution of sample means [23]. This shows the distribution of the means of many random samples of the same size taken from some fixed population. The Z-distribution (shown) represents a standardised version of this distribution (after subtracting its mean and dividing by its standard error). The central region is shown, representing the 95% CI, between the 2.5th and 97.5th centiles of this distribution. The cut-offs at around −2 and +2 enable calculation of the CI. The precise value of these cut-offs depends on sample size and the corresponding degrees of freedom for the T distribution, which corresponds very precisely to the Z-distribution at large sample sizes.

Over many years, I have observed that some students erroneously interpret 95% CIs as having 95% of the data within them, for instance 95% of individual's blood pressure measurements, as reported also by Hoekstra et al. [2]. This error (interpreting 95% CIs as 95% reference ranges) is reflected in some responses to exam questions. I have seen this error at postgraduate level on an MSc in medical statistics, amongst medical students and amongst other student groups. This error might plausibly result from a misreading of the above graph (Fig. 1) because it shows 95% coverage.

My experience (from 2018) is that describing distributions of sample means as "artificial distributions" is useful. I clarified that taking many repeated random samples is not a study design that is used in practice. It is merely a tool that allows us to calculate the width of 95% CIs. This approach does, in my experience, help students to make sense of these distributions. Alternatively, it might at least help them to compartmentalise this information, so that it does not confuse their developing understanding of CIs. This is one strategy that I feel may help steer away from the misinterpretation that 95% of observations lie within the CIs on the mean. There is not, to my knowledge, any published evidence for this approach or reporting how widely this is currently done. I use this alongside other strategies, such as emphasising that CIs reflect precision of estimation, based on random sampling variation. Research into these topics is warranted here, which could include asking students to

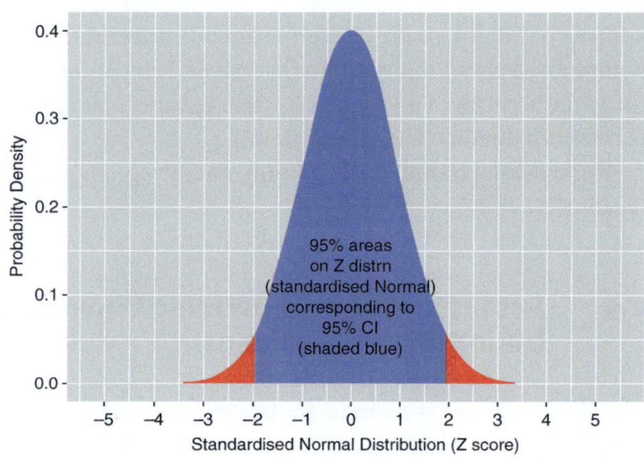

Fig. 1 Distribution of sample means, showing limits for 95% CI

describe their developing understanding. When they gain understanding, they may be able to voice what explanation enabled them to understand and to explain their thought processes.

2.5   Plotting CIs for the Mean on Histograms of the Distribution of Data

In my view, it should be standard practice to show distributions of data, with their CIs, and to emphasise these with interactive exercises. The focus of this section remains on CIs around the mean, which enables this exercise. Such graphs can help students to develop an understanding of CIs. It may further offset the potentially misleading impression given by the above graph showing 95% coverage of sample means.

We provided histograms that show the underlying data, with CIs overlaid onto them (see Fig. 2, used from 2016 and beyond). We asked students to calculate the limits of the same CIs in order to triangulate their learning. By describing CIs as

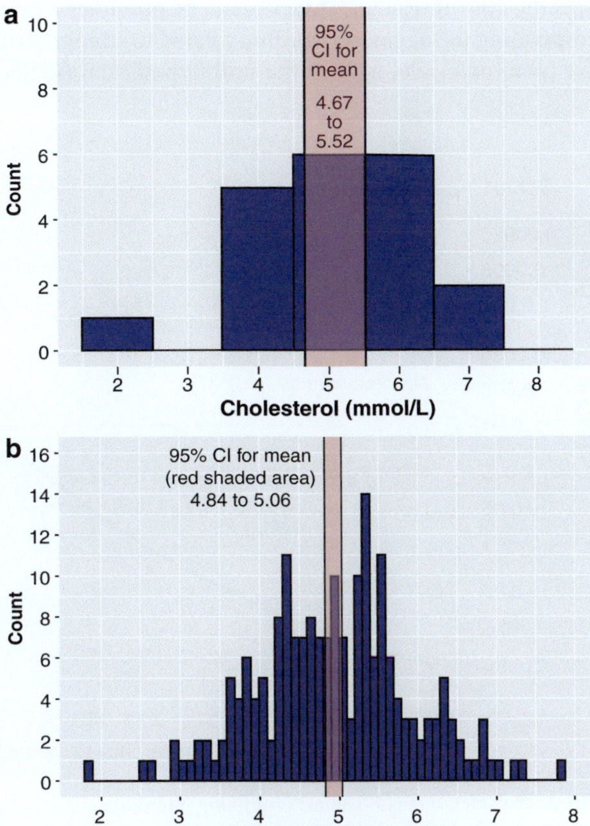

Fig. 2 Histograms showing study participants' blood pressure measurements, with confidence intervals (CIs) overlaid onto them. CIs are shown with dashed vertical lines, and the solid vertical line shows the study participants' mean. (**a**) Sample size 20, (**b**) sample size 200

reflecting precision of estimates, students were readily able to understand the link between the width of CIs (relative to the spread of the distribution) and sample size. This was also demonstrated visually. This certainly helped me later in the course, when occasionally a student forgot what CIs represented. Reminding them of these images was helpful; they recollected the key message from this exercise.

Another exercise that is useful is simply reflecting on the implications of a larger sample size for width of CI, and on varying the standard deviation and the confidence level.

2.6   Promoting the Interpreting of *p*-Values in Terms of Their Strength of Evidence

Interpreting *p*-values by referring to "significance" is of little use in helping students to understand the evidence contained in *p*-values [8, 11]. Referring to results as "statistically significant" arguably implies, in everyday language, that the strength of association is large enough to matter, which may actively confuse. Teaching people to "reject the null" when $p < 0.05$ puts excessive focus on this specific cut-off. Rather, students can be provided with a list of *p*-value cut-offs corresponding to suggested wording related to strengths of evidence for an association (or more precisely, against the null hypothesis) (included in Fig. 3). This use of a

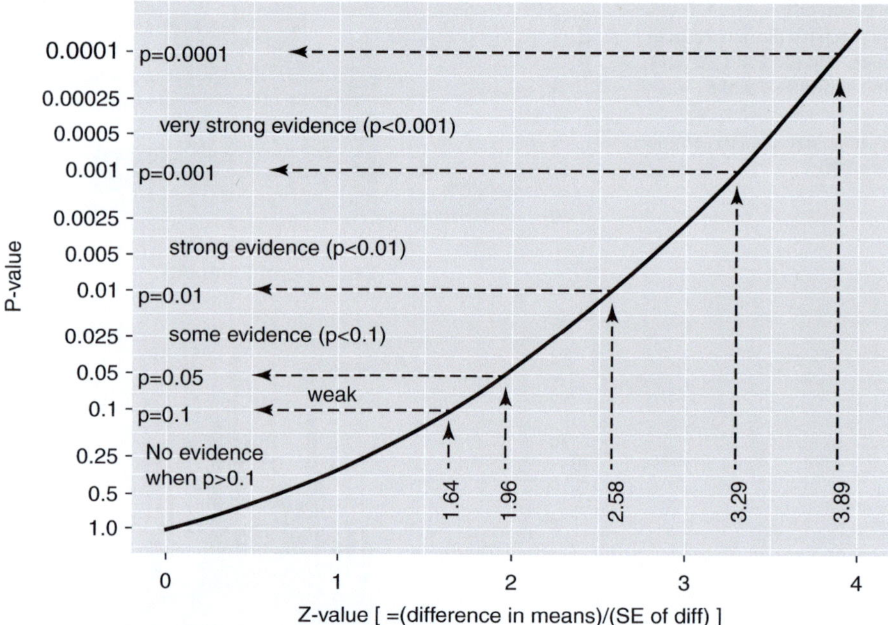

Fig. 3   *P*-values versus Z-values: promoting interpretation of *p*-values as strengths of evidence for associations in the population (assuming random sampling of participants from the population), as appropriate for primary outcome in medical studies. (Figure reproduced with permission [28])

few p-value cut-offs is better than relying on the $p < 0.05$ cut-off alone, and is easy to teach and apply in my experience. We choose to use such interpretations as being a realistic approach on our fast-paced statistics course. However, this is a compromise, since interpreting p-values in terms of strength of evidence, especially evidence for an association rather than against the null, is a step beyond the information incorporated into the p-value. This may be reasonably appropriate for the primary hypotheses because of likely supporting evidence. Caution should be given that these interpretations are not appropriate when p-values are calculated on a speculative basis, perhaps on many variables at a time, with little supporting evidence for the association (including when Bonferroni correction may be considered).

2.7 Promoting the Interpreting of p-Values on a Continuous Scale

The ASA statement on p-values encourages the interpretation of p-values on a continuous scale, rather than looking at them in relation to any specific cut-off [7]. This statement is designed to improve the standard of interpretation of p-values and summarises concerns of that many statisticians have had for many years.

To encourage interpretation on a continuous scale, it is useful to plot p-values (on a log scale) directly against z-values (or t-values for differences in means, for more precise results with small or moderate sample sizes). This shows the relationship between the Z-value/effect size and p-values directly, rather than (as standard practice) via the distribution of sample statistics. The effect size can be described as the ratio of a difference in means to the precision of its estimation (as measured by the standard error). Precision is based on taking the difference in means in study participants to estimate the difference in means amongst the population, assuming random sampling of participants from this population, and dependent on the assumptions underlying these calculations being met.

The Fig. 3 was used in teaching in 2018 and in 2019, to help students to understand the continuous nature of p-values. This was used to find approximate p-values in some student exercises, and as demonstrated by tutors. However, we acknowledge that the addition of "strength of evidence" labelling with associated p-value cut-offs represents the above acknowledged compromise.

2.8 Interpreting p-Values Using Language of Compatibility

Teaching differential interpretation by p-value cut-offs is not in line with the ASA's desired focus on scientific inference, with many different factors being considered, including study design and prior evidence [17]. P-values themselves do not directly provide evidence for associations (or against the null); this is moving towards a Bayesian interpretation, where supporting evidence should appropriately be considered. The calibration between p-values and strength of evidence above is certainly not appropriate for more speculative testing, where there is little supporting evidence for the association being evaluated. The Bonferroni correction [12] might be considered in such situations, although many people including myself consider that to be a rather crude tool.

Using a more direct interpretation of the p-value can encourage higher standards of scientific inference. Hence, teaching material could regularly report p-value definitions, applied to different teaching examples. This is accurate, and yet some may consider it to be lacking in intuitive appeal.

> Example interpreting two-sided $p = 0.002$ against an odds ratio of having asthma in current smokers compared to non-smokers is 1.3: *Assuming no population association, there is a probability of 0.2% that a sample of this size would result in an odds ratio of at least 1.3, or of less than 0.77 (=1/1.3).*
>
> Suggested additions: *(assuming probability random sampling of study participants from this population, dependent on assumptions underlying statistical tests). Considering the "population" is a tool to reflect our interest in generalising results beyond study participants.*

The suggested addition is not conventionally given. It is there to help students to understand the principles underlying p-values. Alternatively, the following p-value interpretation uses the term "compatibility" as advocated by the ASA statement on p-values. It also sticks to the evidence provided directly by the p-value and may have greater intuitive appeal.

> *The data has rather low compatibility ($p = 0.002$) with random selection from a population where smoking and cholesterol are not related (dependent on assumptions underlying the statistical test).*

A figure, analogous to that reported in Fig. 3, can be presented, with "compatibility" wording for the p-value interpretations, which avoids use of fixed p-value cut-offs, as previously published [15]. The ASA promotes the use of the term "compatibility" since this encourages consideration of supporting evidence. Students and researchers can then combine the p-value evidence with other supporting evidence, to come to scientific conclusions [13]. These might be strength of evidence that a certain behaviour is a risk factor for a named disease, or evidence for the effectiveness of a named treatment. This requires an assessment study design and its potential for bias; evidence for bias in practice in the execution of the study, and supporting evidence for mechanisms of action, for epidemiological associations and other treatment studies. Making appropriate scientific conclusions over the strength of the evidence requires a broad range of research expertise and experience. It is a skill that may be developed over time. Whilst students seemed comfortable with the compatibility language, we did not teach scientific inference, at least in part because study design and considerations of bias are taught on a separate module.

2.9   Type I Error Rates and Quantifying Compatibility

This paragraph considers a conventional approach to interpreting p-values, based on the $p < 0.05$ (or other pre-specified) cut-off. This is not intended to advocate such an approach, except to enable sample size calculations. We declare, prior to undertaking research, what p-value cut-off ("alpha") we will use to determine when to reject the null hypothesis. Selecting alpha = 0.05 implies that, conditional on the null hypothesis being true, there is a 5% chance that we reject it (when $p < 0.05$). This "5% chance" is referred to as the type I error rate.

It is tempting to interpret p-values informally as type I error rates, thinking that associations with $p = 0.002$ happen 0.2% of the time (when null hypothesis is true), although this is not a valid interpretation [13]. Such interpretations overlook the fact that this is the probability for the effect size found *or more extreme*. I personally find it tempting to think in these terms, because the definition of the p-value itself can feel cumbersome or contrary to the type of evidence that we would ideally like to see.

Since the ASA p-value statement suggests interpreting p-values using "compatibility", we may naturally want to quantify this compatibility. We have 0.4% compatibility (when $p = 0.002$), for associations which are *on average* this strong amongst study participants (including all associations at least this strong or stronger and an equal quantity of associations that are *less* strong, somewhat *under* the observed odds ratio of 1.3). Hence, we can routinely double the p-value to quantify compatibility, which results in compatibility of associations that are *on average* as strong as those found amongst study participants. This gives an appropriately more cautious approach than informally and erroneously interpreting p-values themselves as type I errors. Quantifying compatibility has the advantage of avoiding the need to change the p-value into some wording relating to strength of evidence, and hence offers encouragement to treat p-values on a continuous scale.

2.10   Biased *p*-Values and CIs (and Biased Odds Ratios or Other Measures) When Analysis Plan Is Not Pre-specified

The final key issue covered by the ASA statement is that p-values are only valid when not selectively reported. Therefore, pre-specified analysis plans are required to avoid bias in p-values and in estimated strengths of association (including in their CIs). When results that are deemed to be most interesting are selectively reported, this selection criteria introduces bias into reported results [29]. We teach MPH students the need for pre-specified analysis plans for clinical trials. Since pre-specification is not always viable, full reporting of all analyses and the basis of decisions to perform additional analyses is considered a reasonable alternative. Alternatively, analyses can be considered as hypothesis generating, requiring testing on external datasets [7]. We teach MPH students to build regression models. To avoid bias, we teach students to choose variables for inclusion based on a literature review and variables required for the research question. We teach that all regression model-building methods that are based on statistical criteria to determine what to adjust for are necessarily biased [30, 31]. When statistical criteria are used to select between many variables in relatively small datasets, the introduced bias may be substantial. However, when statistical criteria are used to select between a small number of variables with very large datasets, bias is usually minimal. We give some guidance on use of statistical criteria, since they are widely used in practice. This includes some guidance on which selection methods are less prone to bias. These considerations were taught to MPH students in 2020 and 2021. Our experience suggests a need to focus on a few core concepts to teach to students and to acknowledge that compromises may be made in practice.

2.11 Student Feedback on MPH Course That Included Many Strategies to Aid Conceptual Understanding

There were several changes to the MPH statistics module each year. This included a more rigorous focus on how interpretation of regression coefficients flowed from regression equations introduced in 2018, and more revision material being made available. This section focuses on feedback from 2018, when many of these teaching methods designed to foster conceptual understanding were introduced into the course. Hence, it relates to the year when comparison with material that did not incorporate these features were most natural and available.

In 2018, graphs in Figs. 1, 2 and 3 were first introduced to develop conceptual understanding of p-values and of CIs. It was the first year that transparent explanatory interpretations of p-values and CIs were used throughout workshop solutions (using slightly different wording [15] to that reported here). Two students described resources that used these techniques as being "awesome". Students earned high marks in a formative multiple-choice test on p-values and CIs in 2018, higher than any of the earlier six student cohorts that took the same formative test (test not used in later years). The exam committee commented on the high number of distinctions in module exams, despite changes to the exam structure such that students were required to demonstrate substantially wider and more sophisticated statistical knowledge than in earlier years. Unlike in earlier years, no students failed. In 2018 and 2019, students had no further statistics teaching prior to their dissertations. Supervisors who routinely supervised a number of students each year reported that these cohorts had especially high levels of statistical ability and confidence. The 2018 student cohort was the first to be tested rigorously in software, adding to students' considerable workload. The senior pastoral tutor (of many years standing in 2018 shortly before she retired) reported especially high stress levels around statistics in this cohort because of their software exam, with many more students than usual coming to her for academic emotional support. Despite this, she noted that this student cohort was notable for the lack of any student coming forward with concerns over a lack of understanding of statistics. She reported that I had earned a reputation for making statistics easy to understand [32].

Independent anecdotal evidence from a university lecturer supports the use of a chosen form of transparent explanatory CIs as a CI "definition" (used routinely throughout a teaching module on Newcastle's master's in public health) as being "enormously helpful" [33], from a university lecturer inspired by my teaching article [15].

Teaching research is warranted to determine which of the included teaching strategies helps students to develop conceptual understanding. Some students may be able to voice what explanation enabled them to understand. They may be able to explain their earlier confusions and their thought processes that lead them to understand, which may illuminate the merits of different strategies.

2.12   Conclusions

Since statistics is considered challenging, it is valuable to consider how we can make inference easier to understand. These teaching ideas are developed from a belief that paying attention to the conceptual understanding of CIs and p-values is important. This includes considering how students might respond to our choice of language. Attention can be paid to known common misunderstandings in terminology and in concepts. We can choose carefully how we interpret/"define" p-values and CIs. Graphs can offer more direct methods to develop understanding of statistical inference, compared to conventional graphs that show distributions of sample statistics. These teaching methods also respond to concerns over poor standards of quantitative interpretation commonly seen in the research literature. Statistics education is believed to be an important factor, hence the focus on methods that are designed to address known common misinterpretations.

These teaching methods describe how teaching of statistical inference can be modified, to comply with the ASA statement on p-values, which is a respected summary of good practice. This encourages greater focus on odds ratios (and similar) amongst study participants and their CIs (or other measures of precision, such as credibility intervals). It discourages the use of any p-value cut-off, promoting the interpretation of p-values as continuous values (if used at all). It supports a focus on scientific inference, rather than the much narrower statistical inference. The introduction of bias into p-values without a pre-specified analysis plan is potentially the most challenging to teach. In many research situations, pre-specification of all analyses is not realistic. Hence practical strategies need to be acknowledged, which are always a compromise.

The strategies presented here may be easier to achieve on a broad scale than the promotion of more complex statistical methods, given the shortage of highly trained statisticians [17]. These basic statistics teaching ideas were developed for a fast-paced 10-day postgraduate course, with modest teaching time for p-values and for CIs. Some of these teaching methods may be appropriate on basic courses on statistical inference at school and basic university level. The desire is that widespread implementation of such teaching approaches would improve standards of interpretation of quantitative results. Teaching research is warranted to obtain evidence for the impact of these methods on student understanding. This may take the form of questioning students over which specific statement or diagram moved their understanding forward.

References

1. Greenland S, et al. Statistical tests, P values, confidence intervals, and power: a guide to misinterpretations. Eur J Epidemiol. 2016;31(4):337–50.
2. Hoekstra R, et al. Robust misinterpretation of confidence intervals. Psychon Bull Rev. 2014;21(5):1157–64.
3. Sotos AE, et al. Students' misconceptions of statistical inference: a review of the empirical evidence from research on statistics education. Educ Res Rev. 2007;2(2):98–113.
4. Steel EA, Liermann M, Guttorp P. Beyond calculations: a course in statistical thinking. Am Stat. 2019;73(sup1):392–401.
5. Tintle N, et al. Combating anti-statistical thinking using simulation-based methods throughout the undergraduate curriculum. Am Stat. 2015;69(4):362–70.
6. Hesterberg TC. What teachers should know about the bootstrap: resampling in the undergraduate statistics curriculum. Am Stat. 2015;69(4):371–86.
7. Wasserstein RL, Lazar NA. The ASA's statement on p-values: context, process, and purpose. Am Stat. 2016;70(2):129–33.
8. Rafi Z, Greenland S. Semantic and cognitive tools to aid statistical science: replace confidence and significance by compatibility and surprise. BMC Med Res Methodol. 2020;20:244. https://doi.org/10.1186/s12874-020-01105-9.
9. Dushoff J, Kain MP, Bolker BM. I can see clearly now: reinterpreting statistical significance. Methods Ecol Evol. 2019;10:756.
10. Haller H, Krauss S. Misinterpretations of significance: a problem students share with their teachers? Methods Psychol Res. 2002;7(1):1–20.
11. Hurlbert SH, Levine RA, Utts J. Coup de Grâce for a tough old bull: "statistically significant" expires. Am Stat. 2019;73(Suppl 1):352–7.
12. Armstrong RA. When to use the Bonferroni correction. Ophthalmic Physiol Opt. 2014;34(5):502–8.
13. Betensky RA. The p-value requires context, not a threshold. Am Stat. 2019;73(sup1):115–7.
14. McShane BB, Gal D. Statistical significance and the dichotomization of evidence. J Am Stat Assoc. 2017;112(519):885–95.
15. Watt HC. Reflection on modern methods: statistics education beyond 'significance': novel plain English interpretations to deepen understanding of statistics and to steer away from misinterpretations. Int J Epidemiol. 2020;49(6):2083–8.
16. Amrhein V, Greenland S, McShane B. Scientists rise up against statistical significance. Nature. 2019;567(7748):305–7.
17. Wasserstein RL, Schirm AL, Lazar NA. Moving to a world beyond "p < 0.05". Am Stat. 2019;73(Suppl 1):1–19.
18. McShane B, Gal D. Blinding us to the obvious? The effect of statistical training on the evaluation of evidence. Manag Sci. 2015;62(6):1707–18.
19. Baker M, Penny D. Is there a reproducibility crisis? Nature. 2016;533(7604):452–4.
20. Lu Y, Henning KSS. Are statisticians cold-blooded bosses? A new perspective on the 'old' concept of statistical population. Teach Stat. 2013;35(1):66–71.
21. Foster C. Confidence trick: the interpretation of confidence intervals. Can J Sci Math Technol Educ. 2014;14(1):23–34.

22. Pfannkuch M, Wild C, Parsonage R. A conceptual pathway to confidence intervals. Int J Math Educ. 2012;44(7):899–911.
23. Kirkwood BR, Sterne JA. Essential medical statistics. 2nd ed. Oxford: Blackwell; 2003.
24. Gelman A, Greenland S. Are confidence intervals better termed "uncertainty intervals"? BMJ. 2019;366:l5381.
25. Kaplan J, Fisher DG, Rogness NT. Lexical ambiguity in statistics: how students use and define the words: association, average, confidence, random and spread. J Stat Educ. 2010;18(2). https://doi.org/10.1080/10691898.2010.11889491.
26. Kaplan JJ, Fisher DG, Rogness NT. Lexical ambiguity in statistics: what do students know about the words association, average, confidence, random and spread? J Stat Educ. 2009;17(3). https://doi.org/10.1080/10691898.2009.11889535.
27. Kaplan JJ, Rogness NT, Fisher DG. Exploiting lexical ambiguity to help students understand the meaning of random. Stat Educ Res J. 2014;13(1):9–24.
28. Watt HC. P-values against Z-values: Interpretation p-values as strengths of evidence for associations in the population, as appropriate for primary outcome in medical studies. 2022 [cited 2022 3/9/2022]. https://www.imperial.ac.uk/people/h.watt/page/innovateteaching.html.
29. Song, F., et al., Dissemination and publication of research findings: an updated review of related biases. Health Technol Assess, 2010. 14(8): 1.
30. Derksen S, Keselman HJ. Backward, forward and stepwise automated subset selection algorithms: frequency of obtaining authentic and noise variables. Br J Math Stat Psychol. 1992;45(2):265–82.
31. Tibshirani R. Regression shrinkage and selection via the lasso. J R Stat Soc Series B. 1996;58(1):267–88.
32. Franey C. Personal communication about Hilary Watt's teaching reputation, based on academic and emotional support offered to struggling students; 2019.
33. Ryan V. Twitter report of experience using Hilary Watt's published (Int J Epi 2020) teaching methods. 2021. 2 August 2022. https://twitter.com/vicky_ryan64/status/1417861494004457480.

Using Directed Acyclic Graphs (DAGs) to Represent the Data Generating Mechanisms of Disease and Healthcare Pathways: A Guide for Educators, Students, Practitioners and Researchers

George T. H. Ellison ⓘD

1 Introduction

1.1 Bridging the Divide Between Clinical and Statistical Expertise Using Directed Acyclic Graphs

The transition from 'knowledge-consumer' to 'knowledge-producer' can require considerable cognitive agility. It can be all the more challenging when the methodological techniques necessary to *produce* knowledge rely on: the prior *consumption* of substantial amounts of very different, and very specialised/technical, types of knowledge; and the confidence of those involved that they have the 'cognitive competence' required to assimilate and apply this [1]. Bridging the divide between disparate forms and levels of expertise is necessary to ensure that advances in any one area can support improvements in understanding and application by *non*-specialists elsewhere; and by generalists and specialists in *several* (or *other*) areas of expertise. Challenges in the building of such bridges exist in many different contexts — not least in the public understanding of advanced scientific techniques (and the 'informed consumption' of the findings, insights and outputs these techniques generate; e.g. [2, 3]). Moreover, many of these challenges are intensifying as a result of the increasing levels of specialism required to attain the levels of knowledge, understanding and technical competence required to break new ground in any given discipline, or on any given topic [4]. For these reasons, the duration and intensity of professional education programmes can struggle to keep up with the rapid pace of advances in knowledge and technology. Nowhere is this more keenly felt than in the fields of statistics, applied mathematics and data science, where a recent step change

G. T. H. Ellison (✉)
Centre for Data Innovation, Faculty of Science & Technology, University of Central Lancashire, Preston, UK

Centre for Academic Technologies, University of Johannesburg, Johannesburg, Gauteng, Republic of South Africa
e-mail: gthellison@uclan.ac.uk

in the volume of deliberately, routinely and incidentally digitised data, and in the computational capacity required to collate, manage and analyse these data, has led to a proliferation of advanced analytical techniques, and a deluge of findings, insights and claims — each of which require expert evaluation and thoughtful consideration prior to their interpretation and application.

For specialists involved in producing knowledge for public consumption — or for use by specialists in *other* domains and by generalists working across *multiple* domains — bridging the divide between the techniques they use, the evidence these generate, and the contribution they might make to strengthening and extending the decisions and practice of others, has led to growing interest in the science of 'exposition', 'translation' and 'implementation'. This has involved the development and evaluation of innovative approaches, techniques and devices designed to facilitate the rapid dissemination, acquisition and integration of novel perspectives; and thereby extend the impact and insight that such perspectives might bring to bear *elsewhere* [4, 5]. These approaches, techniques and devices include: improving the accessibility of specialist or technical language (e.g. [6]; this volume); providing accessible or 'plain language' summaries and 'visual abstracts' of 'key findings' [7]; creating so-called 'user guides' to explicate the 'experiential context' of specialist applications (see for example [8]; and https://mathusersguides.com/); and using a variety of visualisation practices to simplify and demystify the techniques involved and the information these generate [2, 9–11].

Unfortunately, what we know from the use of terms, idioms, similes, analogies and metaphors that cross over from popular (or generalist) to professional (or specialist) 'experiential contexts' is that these risk conflating their very different, context-specific meanings (and very different semantic etymologies) in ways that can mislead, confuse and obfuscate (rather than elucidate, clarify or enlighten). Examples include terms such as: 'significance' and 'prediction', which mean very different things within inferential statistics and popular discourse [12, 13]; and 'machine learning' and 'artificial intelligence', which were ostensibly coined to provide an insight into the complex analytical procedures involved, but have encouraged the widespread popular misconception that both are autonomous, uncontrolled and potentially uncontrollable [14].

Nonetheless, translation appears more successful when the terminology and techniques used are derived from (and remain pertinent, relevant and salient to) the specialist 'producers' involved (as is the case with terms such as 'coefficient' and 'effect size' [15] — although the term 'effect' has itself become a problematised concept within evidence-informed decision-making). Indeed, many of the most successful strategies adopted to-date seem to be those that co-opt a specialist term or device and repurpose this as a medium through which the principles and procedures involved can be more faithfully translated for a wider audience (including the lay public, non-specialists, generalists and specialists in unrelated fields).

This has been the rationale behind the use of directed acyclic graphs (DAGs) to facilitate the acquisition of secondary statistical skills by third year medical students [16–18], since these diagrams help bridge the divide between: the aetiological knowledge and professional/operational experience on which clinical training and practice depend; and the more abstract epistemological and analytical considerations required to extract robust statistical insight from health and healthcare data. DAGs lend themselves to ancillary statistical training precisely because they provide *nonparametric* representations of both: disease and healthcare pathways; and the underlying 'data generating mechanisms' involved — i.e. the mechanisms on which any *parametric* insights generated using inferential statistics depend. As such, DAGs should make inferential statistics much more accessible for those healthcare students, practitioners or researchers who lack high school training or formal qualifications in statistics; and particularly those with little aptitude for, or interest in, the mathematical concepts involved (including those who suffer from 'numerophobia' or 'mathematical anxiety' [19–21]).

1.2   Directed Acyclic Graphs as Epistemological and Educational Tools for Inferential Statistics

Healthcare students, practitioners and researchers unacquainted with DAGs are likely to find them comfortingly familiar and remarkably accessible, since they closely resemble the path diagrams and flowcharts that have become popular for summarising a huge variety of sequential and contingent processes, including disease and healthcare pathways (see, for example: https://www.wikipathways.org/ [22]; and https://cks.nice.org.uk/ [23]; and [24]; see also Fig. 1, below). Indeed, although DAGs are *directed* rather than *directive* — and do not include (or require) the symbols that formal flowcharts use as guides for navigating between 'start' and 'end' points [26] — their use of unidirectional arrows (or 'arcs') to represent causal paths between successive variables (or 'nodes') mirrors the directional 'connectors' that flowcharts use to move from 'start' to 'finish' via a sequence of 'inputs', 'processes' and 'outputs' contingent on 'decisions' made at key points along the way.

Nonetheless, a more critical distinction between DAGs and flowcharts is the former's 'acyclicity'. This requires that no variable can act as its own cause — either *directly* (through a single causal path that starts from, and ends with, the variable concerned) or *indirectly* (through a causal pathway involving a sequence of causal paths between successive variables that eventually loop back to the original variable). Instead, DAGs are inexorably 'progressive' — as befits their temporal disposition — and reflect pathways that link successive pairs of variables over time through a sequence of causal paths between each of the variables acting as a preceding probabilistic cause, and each of the subsequent variables acting as one of their subsequent consequences. It is therefore on the basis of these two key principles — *unidirectionality* and *acyclicity* — that *directed* (and) *acyclic* graphs derive their name; and why DAGs constitute 'principled' diagrammatic representations informed by these two key rules (or 'principles'). More formally, DAGs constitute a subset of causal path diagrams that offer theoretical DAG representations of the known, uncertain or entirely speculative causal relationships thought to be operating within any given context; and thereby offer clear, detailed and principled descriptions of the contributions that theoretical understanding, empirical evidence, or more tentative (or speculative) hypotheses can make to epistemological models of the data generating mechanisms involved [27].

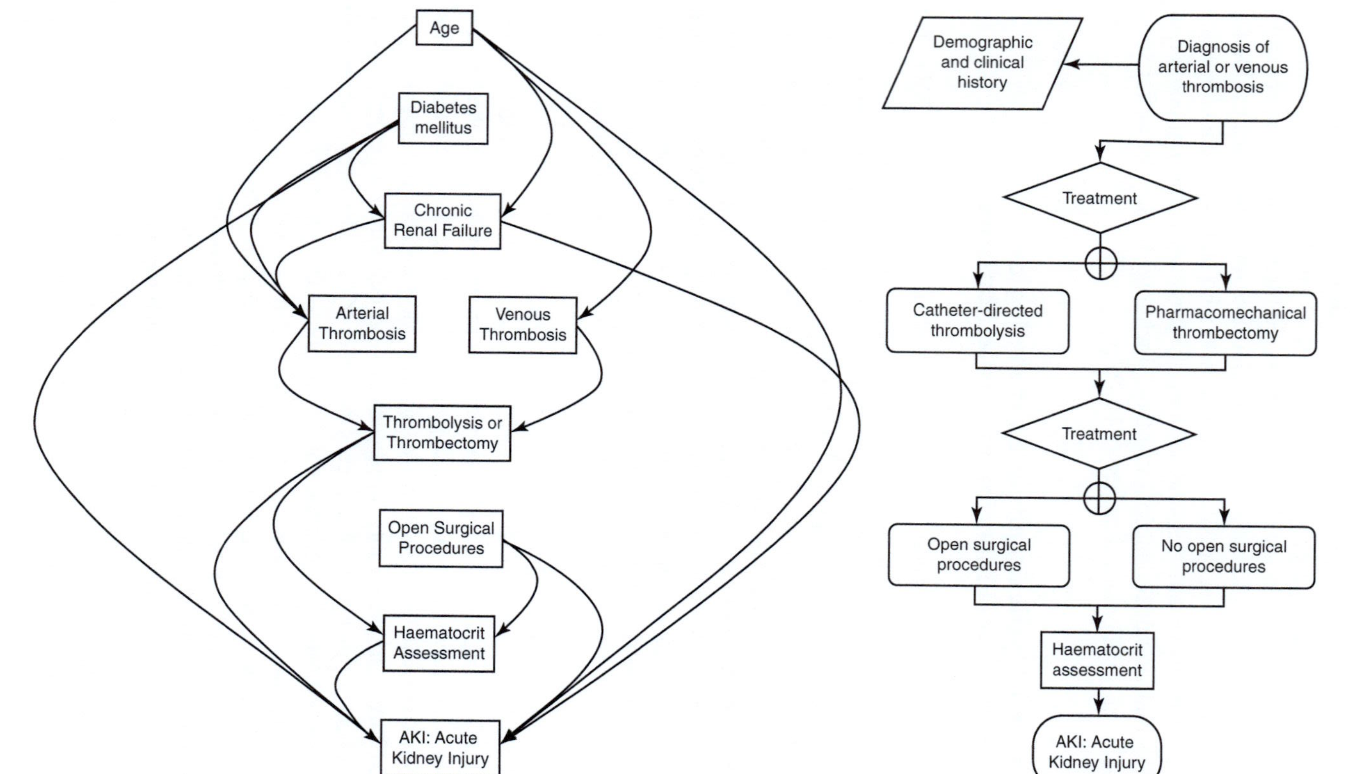

Fig. 1 The similarities and differences between a causal path diagram (below, left), redrawn past-to-present/top-to-bottom from the DAG provided by Escobar et al. ([25]; Figure 7: 244 PA; 29 AAM), to examine the causal effect of two alternative clinical interventions on acute kidney injury in hospitalised thrombosis patients; and a formal flowchart (below, right) drawn to describe the sequence of clinical decisions/outcomes involved in the healthcare these patients received

While DAGs, like flowcharts, only provide *nonparametric* (that is qualitative rather than quantitative) representations of the pathways involved in the processes they describe, both DAGs and flowcharts have *parametric* (i.e. quantitative) implications for the influence that preceding phenomena can exert on subsequent events (be these 'inputs', 'outputs', 'processes' and 'decisions' — as in a flowchart; or, more simply, the variables acting as successive causes and consequences — as in a DAG). This is because the arrangement of causal paths (and in particular, the *omission* of temporally plausible causal paths) imposes strict parametric constraints on the distributional properties of the data sets generated by the mechanistic pathways they represent.

DAGs can therefore be drawn without the need to consider whether the causal pathways involved are strong or weak; positive or negative; precise or diffuse. They can also be drawn with little (if any) knowledge of the: parametric characteristics of the variables concerned; or the distributional properties these pathways impose thereon. Yet DAGs successfully map the parametric *consequences* of the data generating mechanisms they describe, and thereby facilitate the *statistical* analysis of the datasets these mechanisms generate — an invaluable benefit for inexperienced or non-specialist analysts keen to derive robust statistical inference from the underlying causal relationships involved. This is what makes DAGs such powerful translational tools for bridging the divide between aetiological knowledge, professional/operational experience, and the skills required for inferential statistics. Put simply, DAGs can be drawn with little concern for statistical or mathematical expertise but, once drawn, make it much easier to conduct statistical analyses, and acknowledge, mitigate and (potentially) eliminate potential risks of analytical or inferential bias involved therein.

1.3   The Vulnerability of Directed Acyclic Graphs to Imperfect Aetiological Knowledge and Professional/ Operational Experience

Despite the translational benefits of DAGs — which, as we have seen, stem from their dual utility as models of disease and healthcare pathways, and of the underlying data generating mechanisms these entail — drawing DAGs that *accurately* represent the processes involved can be challenging. While one might expect this to be least troublesome for healthcare pathways — since these are ostensibly the artefactual products of (known) operational designs and (deliberate) professional practices — such pathways often depend upon a complex interplay of contextual (structural, social and cultural) factors that can be uncertain, obscure and very different to what was *intended*, and therefore extremely challenging to accurately (or even adequately) describe. Likewise, despite ongoing advances in theoretical understanding, aetiological knowledge and technical expertise, few (if any) disease pathways are *comprehensively* understood. And even where such understanding, knowledge and expertise exist, the processes involved are often contingent on a similarly complex interplay of contextual factors that influence their incidence, presentation, observation, recognition, diagnosis, treatment and prognosis. Such factors can therefore make *both* healthcare *and* disease pathways susceptible to uncertainty and misunderstanding, even by those with advanced training, first-hand experience or specialist expertise. Indeed, the level of uncertainty involved is likely to be higher still amongst healthcare students and less experienced practitioners and researchers whose understanding will depend almost entirely on the extent of the knowledge they have received in their formal training (together with whatever, potentially limited and potentially partial, exposure they have had to the disease and healthcare pathways concerned).

Moreover, even when substantial relevant knowledge, insight and experience of the variables and causal pathways *is* available, any DAGs based thereon will remain susceptible to 'information bias' [28, 29] and 'anchoring bias' [30, 31]. This is because their content and structure will be: constrained by what is likely to be incomplete or inaccurate *information*; or *anchored* to limited prior knowledge or imperfect biomedical/organisational paradigms. DAGs will also be vulnerable to the so-called 'availability heuristic' [32, 33] whenever they (sub)consciously preference information that is most amenable to observation or measurement (such as including only those variables for which data are already, routinely or readily *available*). Under these circumstances, any such DAGs will offer limited scope for statistical insight free from analytical or inferential bias whenever they ignore or overlook the important role that limited information (such as omitted or unmeasured variables) might play therein.

1.4 The Utility of Temporal Logic When Teaching, Learning and Using Directed Acyclic Graphs

Fortunately, the temporal-causal nature of disease and healthcare pathways mean that the DAGs used to represent these (and the underlying data generating mechanisms involved) can often be drawn using temporal logic alone. In fact, provided the temporal sequence of all relevant variables (or, rather, the phenomena, characteristics and features these variables represent) is known or can be inferred with a fair degree of certainty, temporality alone will dictate which causal paths are *impossible* and which might at least be *plausible*. This level of certainty can be achieved simply on the basis that no subsequent variable can *possibly* cause any preceding variable, while any preceding variable can be considered a *plausible* (or so-called 'probabilistic') cause of any subsequent variable, albeit in the absence of *definitive* evidence to the contrary (such as the deliberate, random allocation of preceding characteristics amongst participants who do and do not receive an intervention — as is the case within a randomised controlled trial).

For these reasons, once one has specified a causal pathway of interest (what Aneshensel [34] helpfully called the 'focal relationship') — comprising the plausible, probabilistic causal path between a preceding cause (or 'exposure' variable) and a subsequent consequence (or 'outcome variable') — all that is required to draw a *preliminary* DAG using temporal logic is to:

- **Step 1** — compile a comprehensive list of any phenomena, characteristics or features (and their associated variables) considered relevant to the focal relationship;
- **Step 2** — arrange all of the variables identified at Step 1 in the most plausible temporal sequence on the basis of when the phenomena, characteristics and features involved (as and when each of these were measured) were most likely to have occurred; and
- **Step 3** — include *all* of the plausible, probabilistic causal paths from any preceding variable to all subsequent variables (i.e. as assessed at Step 2, above), only omitting those paths where there is *definitive* evidence these do not or cannot exist.

The aim of this chapter is to illustrate how educators can apply these three (and seven further, simple) steps to strengthen the analytical competence and statistical confidence of healthcare students, practitioners and researchers using DAGs informed by temporal logical. The sections that follow include a worked example involving the critical appraisal of a published clinical study [25] to demonstrate how each of the tasks required to generate a temporally robust DAG can also be used to: evaluate the analytical decisions made during applied healthcare research; and inform the decisions required when selecting the datasets and statistical adjustments required to eliminate, mitigate or (at the very least) acknowledge the risk of bias in similar, real-world, non-randomised studies (i.e. those dependent on routinely or deliberatively collected observational data).

Escobar et al.'s [25] study was chosen as the basis for this chapter's worked example to represent the 144 studies reviewed by Tennant et al. [35], each of which had used (and shared) DAGs to inform the design of their applied health research. Like many of these studies, Escobar et al. [25] drew on a combination of routinely and prospectively collected sociodemographic and clinical data from a discrete subset of patients (in this instance, following a diagnosis for thrombosis), to estimate the strength, direction and precision of a clearly specified focal relationship (in this instance, between the type of treatment received/allocated and a marker of subsequent kidney damage). In common with many of the other studies reviewed, Escobar et al.'s [25] analyses involved multivariable statistical modelling in which the adjustment sets used were informed by a DAG that had been drawn to represent the aetiological and healthcare pathways involved, and determine which variables might act as potential confounders (and should therefore be included in the multivariable adjustment sets used). However, in contrast to some of the other articles that Tennant et al. [35] reviewed, Escobar et al. [25]: contained all of the information required to evaluate the key analytical decisions made; was considered accessible to non-specialists; and included just one clearly drawn (and therefore unambiguous) DAG, containing a manageable number of discrete variables ([25]; Figure 7: page 244 of the published article [PA]; page 29 of the Author Accepted Manuscript [AAM]).

The first three of these steps were used to generate the DAG presented in Fig. 2 (below).

Task 1: Compile a comprehensive list of any phenomena (and their associated variables) considered relevant to the causal pathways of interest:

DoB: Date of birth-determined age at intervention

CRF: Pre-intervention diagnosis of chronic renal failure

IP: Indication for the procedure (venous vs. arterial)

HCT_1: Post-intervention haematocrit values

DM: Pre-intervention diagnosis of diabetes mellitus

INT: Intervention (PPT vs. CDT)

SURG: Post-intervention open surgical procedures

AKI: Acute kidney injury

Task 2: Arrange all variables considered relevant for inclusion in the most plausible temporal sequence on the basis of when the phenomena involved (as, and when, measured) were most likely to have occurred:

| DoB | DM | CRF | IP | INT | SURG | HCT_1 | AKI |

Task 3: Include all of the plausible, probabilistic causal paths between any preceding variable and all subsequent variables (except where there is definitive evidence that no such paths can, or do, exist):

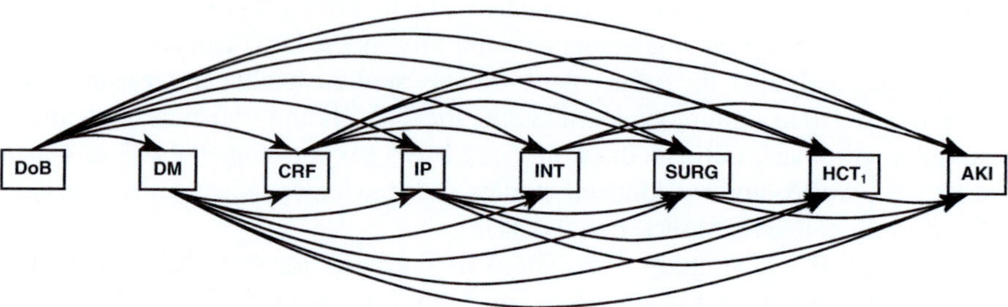

Fig. 2 An illustration of the first three tasks involved in drawing preliminary DAGs based on temporal logic, using the eight discrete variables included in the DAG provided by Escobar et al. ([25]; Figure 7: 244 PA; 29 AAM), to examine the causal effect of percutaneous pharmacomechanical thrombectomy (PPT) vs. catheter-directed thrombolysis (CDT) on acute kidney injury in thrombosis patients. (See: DAGitty)

This comprises a reformatted and redrawn version of the DAG provided by Escobar et al. ([25]: Figure 7: 244 PA; 29 AAM) which was: generated after identifying the eight discrete variables from the nine nodes the authors had included in their original DAG; and based around the focal relationship between INT: the clinical procedure/intervention allocated/received (the specified exposure variable), and AKI: acute kidney injury (the specified outcome variable). At this stage, it is assumed that this list of eight discrete variables comprised those the authors considered both important and sufficient (if not necessarily *comprehensive*) to describe the relevant causal pathways of interest, and the data generating mechanism involved. In this regard, it is worth pointing out that perhaps as many as 75 different phenomena, characteristics, features, and their associated variables were mentioned or cited in the three main sections (Introduction, Methods, Results and Discussion), two Tables (I and II) and seven Figures (1–7) of Escobar et al.'s [25] article. These have been summarised and tabulated in the appendices to this chapter (see Appendix 1), and indicate the far larger number of variables the authors themselves considered relevant to their study's rationale, design, context, sample, analyses and interpretation, yet failed to include (either as measured or unmeasured/latent variables) in the DAG they presented in Figure 7: 244 PA; 29 AAM).

It should be clear from Fig. 2 that these first three, preliminary steps should make drawing (and teaching) DAGs both simple and accessible, even for those who have little if any aetiological/professional knowledge, or first-hand operational experience of the disease and healthcare pathways concerned [16]. Moreover, given the risks of anchoring and information bias, it may not be unreasonable to speculate that drawing such DAGs based on temporal logic *alone* might actually generate *more* objective, and perhaps *more* accurate, representations of the pathways concerned than those drawn on the basis of existing biomedical paradigms or professional/operational understanding (given these are inevitably partial, imperfect or incomplete). Indeed, even when there is a substantial degree of *un*certainty regarding the precise temporal sequence of the variables involved, DAGs informed by temporal logic should still have substantial utility for: prompting epistemological reflection and thoughtfulness; and assessing whether the causal paths considered likely or necessary (on the basis of prevailing knowledge, insight and experience) are consistent with those considered possible (and therefore plausible) on the basis of temporality — something that is remarkably easy to overlook when one is *anchored* to established biomedical/operational paradigms or analytical designs.

2   Drawing Directed Acyclic Graphs Using Temporal Logic

2.1   Challenges Facing the Development of Sufficiently Comprehensive and Temporally Robust DAGs

Nonetheless, while none of the three *preliminary* steps required to generate a DAG on the basis of temporal logic appear particularly taxing, the first two face a number of substantive challenges. This is because compiling a *comprehensive* list of all variables likely to be relevant to any given causal pathway/focal relationship (Step 1), and arranging these in their correct temporal sequence (Step 2), requires very careful consideration of any variables that might be:

- 'omitted' — not least those that are known, measured yet overlooked, but also those that are: known but unmeasured; unknown and therefore unmeasured; and potentially 'unknowable' and therefore unmeasurable and likely to be unacknowledged [36];
- 'time-variant' — and therefore highly dependent on precisely when and how they were measured, and the relevance of the values these measurements provide to the causal effects of their *preceding* values;
- 'non-asynchronous' — such that they are impossible to confidently position before or after one or more other variables, which therefore appear to have occurred at exactly (or essentially) the same point in time; and/or
- 'temporally obscure' — when there is intractable uncertainty regarding their temporal position with regard to one or more of the other relevant variables (and,

in particular, in relation to the specified exposure and outcome variables of the focal relationship of interest).

Such variables can pose substantive challenges when drawing a DAG since: the first (omitted variables) include those that are easily overlooked and, as a result, are commonly under-represented within DAGs (see also Appendix 1); the second (time-variant variables) are difficult to operationalise as discrete, temporally positioned phenomena (not least when these actually relate to characteristics or features rather than 'events' per se); the third (non-asynchronous variables) introduce considerable uncertainty regarding their causal relationships with other simultaneously occurring/non-asynchronous variables; and the fourth (temporally obscure variables) can undermine the whole premise of drawing DAGs using, or informed by, temporal logic [16].

The four sections that follow deal with each of these challenges in turn; offering guidance on how they might be addressed through a series of additional steps which move beyond simplistic, *preliminary* DAGs (such as that summarised in Fig. 2, above), to more comprehensive DAGs that acknowledge (and can therefore help mitigate, or even eliminate), the potential impact of omitted, time-variant, non-asynchronous and temporally obscure variables on the accuracy and analytical utility of the data generating mechanisms such DAGs aim to represent.

2.2 Challenges Posed by the Omission of Measured, Unmeasured and Unacknowledged Variables

Compiling a comprehensive (yet finite and manageable) list of variables considered relevant to any given causal pathway(s) will always require some pragmatic decisions — particularly regarding which of the measured (and therefore available) variables, and which of the unmeasured (and unmeasur*able*) variables, might be both *necessary* and *sufficient* to represent the data generating mechanism(s) involved (see also Appendix 1). This is evident from Tennant et al.'s [35] recent review of 144 published DAGs, in which the number of variables included ranged from 3 to 28 with an average of just 12. This may partly reflect the decline in readability that accompanies the inclusion of more than a dozen variables (and their associated causal paths) within a static two-dimensional DAG [16]. However, it also seems likely that (for the most part) the specific variables selected for inclusion in many of these DAGs were either: those for which measurements were readily or already available; or those considered critical to include on the basis of prior aetiological knowledge (both theoretical and empirical) or professional/operational experience. If so, then the very basis upon which many contemporary DAGs are drawn will be vulnerable to the 'availability heuristic' and 'anchoring' bias, respectively.

Notwithstanding the need to recognise, reduce and (where possible) avoid the risks that such cognitive biases entail [37], it remains common practice to ignore unmeasured variables when drawing DAGs. Indeed, Tennant et al. [35] reported that only a third of the 144 DAGs they reviewed contained one (or more) unmeasured variable(s), while none appeared to have included *unacknowledged* variables — i.e. the 'unknown unknowns' and potentially '*unknowable* unknowns' that fall outside existing knowledge, theory or perception, and are therefore unmeasurable [36]. A failure to consider the potential relevance of such variables when drawing a DAG both replicates and reinforces the narrow gaze imposed by excessive reliance on: (partial) aetiological knowledge; (limited) professional/operational experience; (imperfect and incomplete) biomedical and organisational paradigms; and variables for which measurements are more readily (or are already) available. When no unmeasured variables and no unacknowledged variables are included, the DAGs drawn are much more likely to misrepresent the complexity of the causal pathways involved. They also overlook: the important role(s) that any of the omitted variables (whether known, unknown or unknowable) might play therein; and the potential utility of conceptualising and operationalising such variables within the theoretical and empirical deliberations that DAGs are intended to promote.

For these reasons, all such variables — including those that are measured, unmeasured *and* unacknowledged — should be considered relevant to include when drawing a DAG since all of these can: prompt reflection and generate insight regarding alternative aetiological and operational pathways; and help highlight potential sources of analytical and inferential bias that might otherwise be overlooked. To this end, any 'measured-yet-omitted' and all 'known-but-unmeasured' variables should be positioned wherever they are likely to have occurred within the temporal sequence of variables already selected for inclusion; while hypothetical (multivariable) 'sets' of unacknowledged variables should be distributed throughout the temporal sequence of (measured and unmeasured) variables, and particularly at those points in time where they would pose specific and credible risks of analytical and inferential bias to any subsequent statistical analyses (as described in more detail in Sect. 3.2, below).

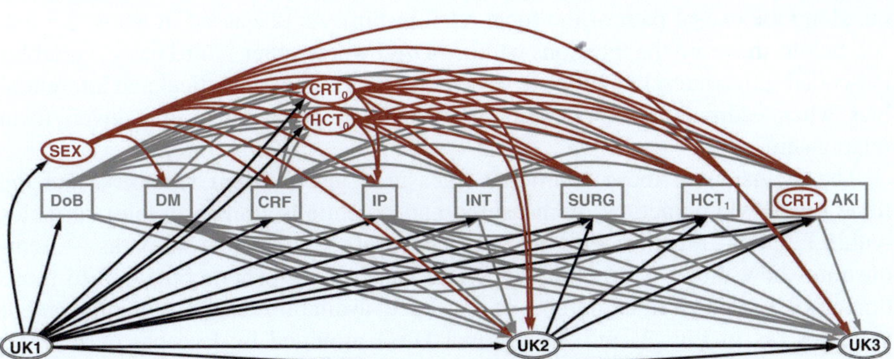

Fig. 3 A further iteration of the DAG presented in Fig. 2 to illustrate how: (1) additional, measured-yet-omitted and known-but-unmeasured/unreported variables (SEX, CRT_0, HCT_0 and CRT_1 — in **red font**); and (2) unknown and hitherto unacknowledged sets of variables (UnKnown 1-3: UK1, UK2, and UK3 — in **black font**) might be added to prompt reflection and generate insight regarding: alternative aetiological/operational pathways; or potential sources of analytical/inferential bias that might otherwise be overlooked; and (3) how variables considered non-asynchronous (such as: DoB and SEX; and CRT_0 and HCT_0) might best be located (as events occurring simultaneously with no causal pathways operating between them). (See: DAGitty)

To illustrate how unmeasured and unacknowledged variables might be usefully incorporated into a DAG, Fig. 3 extends the worked example (Fig. 2 — based on the DAG provided by Escobar et al. [25]: Figure 7: 244 PA; 29 AAM) to reflect two further steps:

- **Step 4** — include all relevant variables (including any measured-yet-omitted, and all known-but-unmeasured, variables); and
- **Step 5** — include hypothetical sets of potentially unacknowledged (and therefore unknown, unmeasured and potentially unmeasurable) variables at key points within the DAG to prompt consideration and reflection regarding the risks of analytical and inferential bias these might pose.

The first of these steps (Step 4) identified three additional measured variables (the sex of the patients examined — SEX; their baseline creatinine levels — CRT_0; and their baseline haematocrit values — HCT_0), all of which had been reported in Table 1 of Escobar et al. (25), yet all of which were omitted from the published DAG and the multivariable analyses the DAG was designed to inform (para. 1: 241 PA; 21 AAM). It also identified a fourth measured, but unreported (and therefore omitted), variable, namely: the follow-up creatinine values of the patients examined (CRT_1; recorded within 72 hours of INT) which, though omitted from the DAG and multivariable analyses, nonetheless informed (and were integral to) the diagnosis of AKI.

Meanwhile, the second of these steps (Step 5) involved: positioning three hypothetical sets of variables (UnKnown 1-3: UK1; UK2; and UK3): prior to the very first of the measured variables (the date of birth-determined age of the thrombosis patients; DoB); after the very last of the measured variables (acute kidney injury: AKI); and in-between the specified exposure (INT) and specified outcome (AKI; i.e. along the causal path of the focal relationship). As discussed in Sects. 2.5 and 3.2, below, these are the locations within a DAG where other, 'third party' variables (so-called 'covariates') might pose very particular risks of analytical and inferential bias when estimating the strength, direction and precision of any given focal relationship.

The inclusion of these additional (measured yet omitted) variables, and the three sets of (hypothetical, unmeasured and potentially unmeasurable) variables within Fig. 3 — together with their associated probabilistic causal paths — demonstrates how much more complex DAGs can become when compared to those drawn solely using a small number of measured, available, accessible and/or salient variables (as in Fig. 2, and the original DAG provided by Escobar et al. [25]: Figure 7: 244 PA; 29 AAM). This added complexity nonetheless has substantial utility for: tempering excessive confidence that any such models might comprehensively represent the multitude of aetiological and operational pathway(s) involved, and the data generating mechanism(s) these reflect; as well as revealing and exploring alternative target variables and causal pathways that might warrant further investigation and statistical modelling as possible targets for subsequent intervention. At the same time, the inclusion of additional unmeasured and unacknowledged variables can also be invaluable for considering, evaluating and identifying potential sources of analytical or inferential bias, so that these are not ignored or overlooked, and can instead be eliminated, mitigated or (at the very least) acknowledged (see Sect. 3.2, below).

2.3 Challenges Facing the Temporal Positioning of Time-Variant Variables

Notwithstanding the potential benefits of including unmeasured variables within any given DAG, their inclusion is nonetheless predicated upon the ease and accuracy with which the temporal sequence of these (and other, measured) variables can be determined. Likewise, adding hypothetical sets of unacknowledged variables at those points in the temporal sequence where they might best prompt consideration of, and reflection on, the substantive risk of analytical and inferential bias that such variables might then pose is also predicated upon establishing a temporal sequence of measured and unmeasured variables amongst which these additional sets of variables might then be positioned. In practice, determining (or inferring) the temporal sequence of any group of variables will largely depend upon whether the variables concerned represent phenomena, features or characteristics that are 'time-invariant' or 'time-variant' — that is, whether these variables refer to discrete, one-off events, happenings or occurrences (hence 'time-*in*variant'); or are susceptible to change over time (hence 'time-*variant*').

It should be relatively straightforward to identify when a time-invariant variable occurs (or occurred) since this can be traced to the associated event concerned. However, wherever there is substantive variation in the precise point in time at which such events occur these variables can also present as essentially time-*variant* at the population level, making it necessary to estimate the most likely or commonest point in time *at* which (or *by* which) the phenomena concerned will have occurred for *most* (or *all*) of the cases examined (be these study participants or other entities). In a similar fashion, identifying when a time-variant variable occurs (or occurred) is likely to prove much more challenging, because although their values are dependent upon *when* and *how* they were measured, the relevance of these values to the causal mechanisms involved may actually depend upon values attained much *earlier*. Knowing precisely when 'causally relevant' values occur would require repeated measurements, if not regular or continuous monitoring, of all time-variant variables — something that is only rarely undertaken, might often be impracticable (and would potentially be impossible for many of the phenomena, characteristics and features involved in disease and healthcare pathways). For these reasons (and for both time-*in*variant and time-*variant* variables), healthcare students, practitioners and researchers must carefully balance whatever information they have regarding: *when* the variables concerned were measured; *how* these measurements were made; and *what* is known about the nature of the phenomena, features and characteristics concerned.

Only then might it be possible to confidently assess when — and therefore where, with respect to all other variables included in the DAG — these variables might be most appropriately positioned.

In this regard it is helpful to be able to rely on the fact that measurements made at any given point in time can *only ever* relate to phenomena, features and characteristics that have *already* occurred — i.e. *prior* to or *at* the point in time at which the measurements concerned were made. However, the certainty this provides is often only helpful for those variables that are unlikely to have occurred: over protracted periods of time; and, in particular, during periods of time in which one or more of the *other* relevant variables might have *also* occurred. For this reason, even time-invariant variables may only be amenable to discrete temporal positioning where the period of time over which these might feasibly have occurred is distinct from those of other relevant variables (including other time-*in*variant and time-*variant* variables).

Under these circumstances, determining the temporal sequence of variables considered relevant for inclusion in a DAG often relies far more on *how* these variables were measured than on *when* these measurements were made. Since the meaning of any given variable is ultimately determined by the technique(s) involved in its measurement, and since knowledge of these technique(s) is usually available to those making use of such variables (even if only as encoded in the formatting of data provided second- or third-hand), this information can offer valuable insights on which healthcare students, practitioners and researchers can attempt to infer *when* the phenomena, characteristics or features concerned are most likely to have occurred.

For example, measurements that depend upon *preceding* phenomena (or the *preceding* values of other characteristics and features) make it possible to position the variables concerned in an appropriate temporal sequence with respect to one another, as is the case: with many of the identity-, age-, growth-, development- and exposure-dependent biological and social processes involved in lifecourse-related disease pathways; and (perhaps) with most (if not all) of the sector-, structure-, resource- and context-dependent operational procedures and processes that exist within healthcare pathways and the delivery of most other services (see, for example, the DAGs in Ellison and De Wet [38]: 259; and Ellison et al. [39]: 3). Likewise, the techniques required to measure time-variant phenomena can often provide substantive clues as to the period of time (or particular point in time) to which any measured value might relate. This is certainly the case with: disease-relevant biological, social and environmental phenomena that vary or change in a more or less coherent fashion over time, and are likely to have been initiated or determined some time before they 'crystallised' in their measured form; and the levels of expertise, efficiency and performance operating within healthcare systems that might only be achieved as a result of decisions made, or systems that evolved and matured, *prior* to their examination and measurement.

These then are the considerations that lie behind two further steps considered helpful for improving the specification of DAGs:

- **Step 6** — identify which of the included *time-invariant* variables are likely to exhibit substantive variation in the precise point in time at which they occur, and carefully (re)position these at the point in time at which (or by which) the phenomena concerned are likely to have occurred for most (or all) of the cases examined (be these study participants or other entities); and
- **Step 7** — carefully consider whether the measured values of *time-variant* variables are likely to have actually occurred or crystallised sometime before their measurements were made and, where necessary, move the variables concerned to an appropriate, earlier position in the sequence of variables included in the DAG.

For the example summarised in Fig. 3 (above), eight of the measured variables were considered time-invariant: DoB; SEX; DM; CRF; IP; INT; SURG; and AKI. Three of these are events (DoB; INT; and AKI) that demarcate discrete periods of the disease and healthcare pathway, and with respect to which all of the other relevant variables can then be positioned. One other variable (SEX) also constitutes a discrete event, but is determined at the point of conception (subject to later confirmation and formal assignment at the point of birth). As such, these four time-invariant variables do not display any relevant temporal variability when estimating the strength, direction and precision of the focal relationship (between INT and AKI) within the context of Escobar et al.'s [25] study.

Of the remaining four time-invariant variables (DM, CRF, IP and SURG), all are determined by aetiological processes, as reflected in subsequent diagnostic assessments (or therapeutic decisions), which might therefore have feasibly occurred (or been made) at somewhat variable points in time between DoB/SEX and INT (where DM, CRF and IP must have occurred) or between INT and AKI (where SURG must have occurred). When generating the DAG summarised in Fig. 3, it was acknowledged that these events might have all preceded the points in time when the clinical assessments and decisions concerned were made. However, it was assumed that the emergence of DM was most likely to have preceded that of CRF, and that the thromboses on which IP was based were unlikely to have occurred before either DM or CRF — assumptions that might well be at odds with contemporary understanding or empirical evidence concerning the aetiological mechanisms involved.

Unfortunately, Escobar et al.'s [25] article, like the vast majority of published accounts of empirical research, does not (have space to) provide the evidence required to assess (let alone confirm or reject) any of these assumptions. The temporal positioning of these variables was therefore informed as much by speculation (regarding the disease and healthcare pathways thought likely to have been involved) as by the limited clues available from the methods used to measure each variable. Much as these clues (and the modest contribution that the timing of measurements) can help in elucidating the likely temporal sequence of variables considered relevant for inclusion in a DAG; such clues are rarely definitive or incontestable. Drawing DAGs on the basis of temporal logic is therefore as much an 'art' as a 'science', and ultimately depends upon the insight, deliberation, resolute objectivity, temporal perspective and judgement of the healthcare students, practitioners and researchers concerned. And while a temporal approach can add substantial value to the utility of DAGs (not least in the robust test that such DAGs can then provide to more partial professional/operational knowledge and the imperfect paradigms that reflect our incomplete understanding of biomedical and organisational pathways), it often requires diligent reflection, robust debate and a willingness to disclose any substantive uncertainty by offering a range of DAGs describing the alternative temporal sequences considered plausible (see for example [40]; Figure 1: 843).

Meanwhile, of the remaining measured and unmeasured variables included in Fig. 3, four comprise baseline and follow-up measurements of two time-variant variables (HCT_0/HCT_1 and CRT_0/CRT_1), one of which (CRT_1) was integral to (if not essentially synonymous with) the diagnosis of AKI — a variable previously interpreted as time-invariant (given it reflects a diagnostic decision regarding the subsequent health of the patients examined). Indeed, it was for this reason that AKI was considered central to the temporal framework of the study around which all other variables might then be positioned, even though it was measured/diagnosed on the basis of what must have been the known (yet unreported) time-variant follow-up measurement of CRT (hence CRT_1 [25]: 240 PA; 7 AAM). While it is therefore a moot point whether the value of CRT_1 might have *actually* occurred much earlier than the formal diagnosis of AKI, there is little value in considering this possibility since AKI is based upon (and therefore essentially the *same as*) CRT_1. However, it remains possible (and potentially troublesome) that the value of CRT_1 (and its role in the diagnosis of AKI) might have *actually* been achieved *before* HCT_1 or SURG — a possibility with consequences that will be discussed in Sect. 2.5, and at Step 10 (below).

Setting this possibility aside, it nonetheless remains entirely plausible that the values of HCT_0, CRT_0 and HCT_1 may all have been achieved prior to the points at which these were actually measured; and, moreover, that the *measurements* of HCT_0 and CRT_0 may have also occurred over a range of different points in time given the study appears to have drawn on data collected somewhat opportunistically during prior clinical care (rather than at predetermined points in time within a carefully monitored disease and healthcare pathway — as might have been the case were the authors to have adopted a more deliberative prospective study design). For this reason, the development of alternative DAGs (as discussed earlier, and in Sect. 2.4 and at Step 9, below) should also consider the possibility that the values of HCT_0 and CRT_0 were measured and/or occurred *before* CRF or DM.

These considerations have been summarised in Fig. 4 in which the potential, alternative temporal positions of the relevant covariates have been highlighted using coloured boxes spanning the periods in time over which these covariates might plausibly have occurred (or 'crystallised'). These reflect substantial uncertainty regarding the *precise* temporal positioning of: DM, CRF, CRT_0 and HCT_0 with respect to one another (although all four covariates are considered likely to have occurred/crystallised *after* DoB/SEX and *before* IP); and SURG, HCT_1 and CRT_1/ AKI — both with respect to one another *and* regarding the possibility that the value

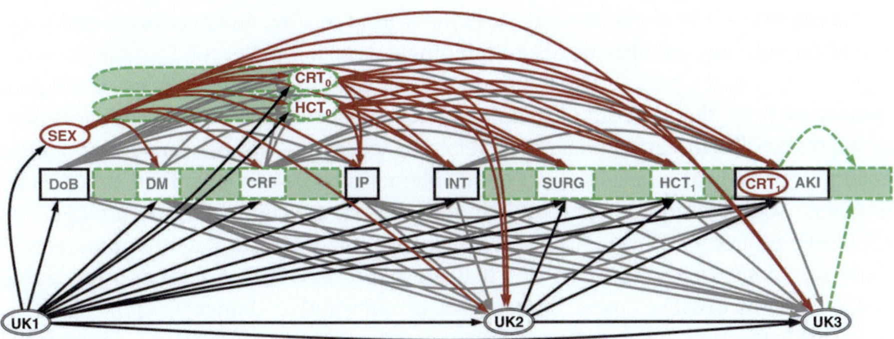

Fig. 4 A further iteration of the DAG presented in Fig. 3 in which the potential, alternative temporal positions of six time-invariant/variant covariates (DM, CRF, CRT_0, HCT_0, SURG, and HCT_1) have been highlighted using boxes (in **green font**) spanning the periods over which these covariates might have plausibly occurred (or crystallised). Were SURG and/or the value of HCT_1 to have occurred/crystallised *after* the value of CRT_1 had crystallised (as and when the values of these variables were measured), additional causal paths might then be required to reflect the plausible probabilistic impact *of* CRT_1/AKI and UK3 *on* SURG and/or HCT_1 (as indicated by the dashed arrows in **green font**). **UK1:** UnKnown/unacknowledged variable set 1; **SEX:** Sex; **DoB:** Date of birth-determined age at intervention; **DM:** Pre-intervention diagnosis of diabetes mellitus; **CRF:** Pre-intervention diagnosis of chronic renal failure; **HCT_0:** Baseline/pre-intervention haematocrit values; **CRT_0:** Baseline/pre-intervention creatinine values; **IP:** Indication for the procedure (venous vs. arterial); **UK2:** UnKnown/unacknowledged variable set 2; **INT:** Intervention (PPT vs. CDT); **SURG:** Post-intervention open surgical procedures; **HCT_1:** Post-intervention haematocrit values; **CRT_1:** Post-intervention creatinine values; **AKI:** Acute kidney injury; **UK3:** UnKnown/unacknowledged variable set 3

of CRT_1 (as and when measured, together with the diagnosis of AKI that these values then permitted) may have plausibly occurred/crystallised *before* SURG occurred and/or *before* the value of HCT_1 (as and when measured) had crystallised (i.e. closer to the point at which this was measured, assuming CRT_1 and HCT_1 were measured at the same time).

2.4   Challenges Facing the Specification of Causal Relationships Amongst Non-asynchronous Variables

Alongside the additional steps required to accommodate omitted and time-variant variables, DAGs also need to be able to deal with subsets of variables that occur or crystallise at (more or less) the same point in time — i.e. those that are 'non-*a*synchronous'. For these groups of variables it is not possible to determine which might act as the cause or consequence of the other(s). It is therefore inappropriate to include causal paths in either direction between non-asynchronous variables within a DAG, even though all such variables are likely to share any preceding (measured, unmeasured and unacknowledged) probabilistic causes and all subsequent probabilistic consequences (notwithstanding definitive evidence to the contrary; see Step 3, above).

Examples of non-asynchronous variables within disease and healthcare pathways might include any number of discrete biological, social or circumstantial phenomena, features and characteristics that emerge as multiple consequences of singular events, such as: the plethora of genotypic characteristics determined at conception; or the metabolic, dietary- and activity-related phenotypes that emerge and crystallise at similar periods of time as a result of the interaction between an individual's genetic heritage, sociocultural background/circumstances, educational opportunities and economic trajectory. Within healthcare pathways, there are similar structural, procedural and contextual characteristics that are likely to: share a common preceding cause (such as dis/investment in resources or substantive changes in policies, contexts, infrastructure, services or facilities); and then lead to simultaneous changes in a wide variety of subsequent phenomena, characteristics and features. When drawing DAGs informed by temporal logic, it is therefore important to: carefully consider whether one or more of the variables included might need to be situated at *essentially* the same point in time within the hypothesised temporal sequence; and take care not to include causal paths between any of these non-asynchronous variables.

However, the decision to position variables at *essentially* the same point in time along the temporal sequence of probabilistic causes and consequences should ordinarily reflect the likelihood that the phenomena, characteristics and features concerned *actually* occurred or crystallised at more or less the same time. Indeed, it would be poor practice to treat variables as non-asynchronous simply whenever there is uncertainty regarding which of these might precede the other. Instead, such uncertainty is best addressed by offering a range of DAGs describing all of the alternative temporal sequences considered plausible (as suggested in Sect. 2.3, above).

These considerations are encompassed in two further steps that can help improve the accuracy and functional utility of DAGs:

- **Step 8** — ensure there are no causal paths between any variables considered non-asynchronous; and
- **Step 9** — wherever substantive uncertainty prevails regarding the temporal position of any variables (following *any* of the preceding steps), provide a comprehensive range of alternative DAGs in which the variables concerned occupy alternative, plausible temporal positions in the sequence of variables included.

For example, when adding four measured yet omitted covariates to the DAG in Fig. 3 (these being: patient sex: SEX; baseline creatinine and haematocrit values: CRT_0 and HCT_0; and follow-up creatinine values: CRT_1), it was judged that the first of these (SEX — a time-*invariant* variable) would have occurred at *essentially* the very same time that each of the patients included in the study were conceived or born (i.e. DoB). In contrast, the second and third of these (CRT_0 and HCT_0 — both of which were time-*variant* variables) were considered likely to have crystallised: over a more diffuse period of time (i.e. as and when measured during the period of prior clinical care encompassed by this study); and at approximately, and therefore *essentially*, the same point in time with regard to one another. This was the rationale behind positioning both of these variables *after* the two time-*invariant* markers of pre-thrombotic chronic disease (diabetes mellitus: DM; and chronic renal failure: CRF) and *before* the onset of thrombosis (whether arterial or venous: IP). Likewise, given the dependence of the specified outcome (AKI) on the (known, measured, yet unreported and initially omitted) follow-up measurement of creatinine (CRT_1), these two variables were considered synonymous and therefore *entirely* (rather than *essentially*) co-terminous.

Were more frequent measures of CRT and HCT available (both before and after the point at which CRT_0 and HCT_0 were *actually* measured), it might have been plausible to select values from amongst these measurements so as to be able to position the two covariates asynchronously within the DAG, with one acting as a plausible probabilistic cause of the other. But from a purely temporal perspective — and in terms of the causal roles these covariates play with regard to the specified exposure variable (INT) and the specified outcome variable (AKI) — the decision to situate CRT_0 and HCT_0 at (*essentially*) the same point in time within the DAG seems entirely appropriate, albeit somewhat pragmatic.

On the basis of these assessments, any alternative DAGs drawn to explore intractable uncertainty regarding the relative position of time-*in*variant variables (particularly those that might have occurred over a substantial period of time), and time-*variant* variables (particularly those whose values, as measured, may have occurred sometime before they were measured), would involve exploring both: the sequence of variables occurring between DoB/SEX and INT (i.e. DM, CRF, CRT_0, HCT_0 and IP); and those occurring between INT and CRT_1/AKI (i.e. SURG, and HCT_1). However, it is important to point out that the number of additional DAGs required to accommodate variation in the temporal positioning of these covariates (5 between DoB/SEX and INT; and 2 between INT and CRT_1/AKI) would be no fewer than 240, and many more were it necessary to accommodate all possible permutations of non-asynchronously occurring pairs or groups of variables therein. As such, Fig. 4 (which seeks to accommodate the plausible, alternative temporal positions of DM, CRF, CRT_0 and HCT_0); and SURG, HCT_1 and CRT_1/AKI) conceals the very large number of possible permutations (and alternative DAGs) involved.

2.5   Challenges Posed by Intractable Uncertainties Regarding Temporally Obscure Variables

For this reason, although Step 9 can help to accommodate and explore any intractable uncertainty regarding the likely temporal positioning of measured and unmeasured variables, it achieves this only at the cost of generating additional (and potentially very large numbers of different) DAGs. Moreover, the sizeable number of alternative DAGs that are required to accommodate even a modest degree of temporal uncertainty (as was the case with the worked example presented here) is likely to offer little more than pause for reflection wherever these DAGs are intended to support more practical or pragmatic applications (such as: expert or peer review of the theorised pathways they aim to represent; or robust statistical analysis of the underlying data generating mechanisms involved). At the same time, drawing DAGs that depend upon the (potentially sparse) temporal positioning clues available — from whatever techniques were used to measure each of the time-variant and time-invariant variables included — risks relying once more on prior knowledge, insight or experience and (re)introducing the very same cognitive biases that drawing DAGs using temporal logic sought to avoid. Such risks therefore appear inevitable and may only, at best, be subject to a modest degree of mitigation by focussing intently on temporality.

Nevertheless, in most comparable scenarios to that described by Escobar et al. [25], the need to accurately determine the relative temporal position of *every* variable considered relevant to include in any given DAG may not be *strictly* necessary. This is certainly the case for any hypothesised set(s) of unacknowledged variables, whose temporal position(s) are chosen by the healthcare student, practitioner or researcher concerned, and *simply* to reflect where such variable sets might pose potential risks of analytical and inferential bias. This is also the case whenever the intended application of the DAG is solely (or primarily) to acknowledge or reduce the risk of bias when estimating the strength, direction and precision of a *single* causal path (i.e. a single focal relationship) rather than developing a more comprehensive representation of any given data generating mechanism from which the risk of bias might then be addressed for *multiple* causal paths/focal relationships.

Indeed, the benefit of focussing on just one focal relationship when drawing a DAG is evident from the worked example presented in this chapter. This is because the stated aim of Escobar et al.'s study ([25]: 239 PA; 6 AAM) was principally: 'to determine if patients undergoing AJ [AngioJet™] thrombolysis were at increased risk of acute renal injury, when compared with non-AJ CDT [catheter-directed thrombolysis] …'. As such, their study focussed intently on estimating a single focal relationship between one specified exposure variable (INT: the clinical procedure/intervention allocated) and one specified outcome variable (AKI: acute kidney injury — defined as 'an increase in Cr [creatinine values] >25% of baseline within 72 hours [of the intervention], or an absolute increase of >0.5 mg/dL' [25]: 240 PA; 7 AAM).

With only a single focal relationship involved, the temporal position of all other variables considered relevant to include in the DAG inevitably falls into one of just three discrete periods of time: before the specified exposure variable; after the specified outcome variable; or somewhere between the two. Crucially, it is these three periods (rather than the specific sequence of variables within each of these periods) that determines: whether any of the other variables (i.e. the covariates) might act as plausible probabilistic causes or consequences of the specified exposure and/or outcome variables; and which of these might then pose a risk of conditioning-dependent analytical and inferential bias when estimating the strength, direction and precision of the focal relationship (as discussed in Sect. 3.2, below). For this reason — and whenever the principal aim of generating a DAG is to eliminate, mitigate or simply acknowledge the risk of bias when estimating a single focal relationship — provided it is possible to determine whether each of the covariates occurred or crystallised before, after or in-between the specified exposure and outcome variables concerned, it is rarely necessary to correctly determine the respective temporal position of those covariates occurring *within* any of these three distinct periods of time.

There is therefore a final step that is well worth considering to simplify the representation of causal pathways within a DAG, particularly when these pathways face intractable uncertainty as a result of temporally obscure variables (including those variables likely to have occurred or crystalised over protracted periods of time)

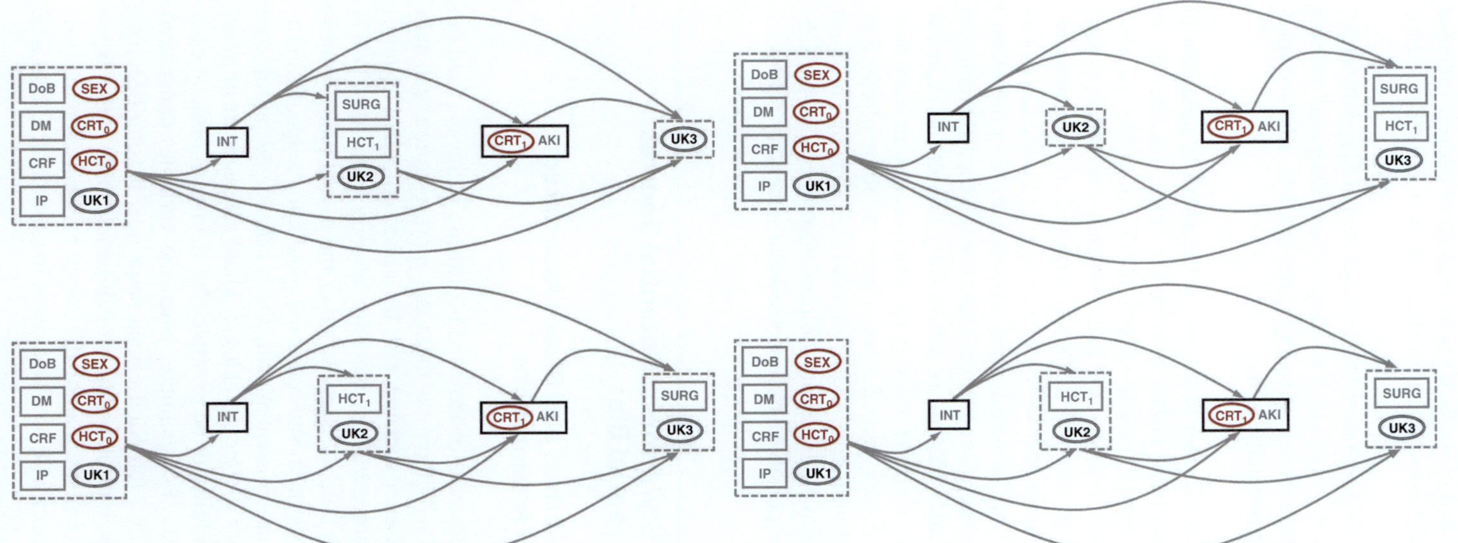

Fig. 5 Four simplifications of the 240+ alternative versions of the DAG presented in Fig. 3 (redrawn from the DAG provided by Escobar et al. [25]: 244; Figure 7: 244 PA; 29 AAM) required to accommodate residual uncertainty in the temporal positioning of two key covariates (SURG and HCT_1) in which each of the measured (**grey font**), unmeasured (**red font**) and unacknowledged (**black font**) covariates has been included in one of three discrete sets of variables, enclosed within separate boxes and located either: before the specified exposure variable (INT); after the specified outcome variable (CRT_1/AKI); or in-between the two. This approach to simplifying DAGs clearly reduces the very large number of alternative DAGs required to accommodate uncertainty in the temporal positioning of covariates therein, while retaining their utility for identifying those covariates that present a potential risk of conditioning-dependent analytical and inferential bias when estimating the strength, direction and precision of the specified focal relationship (in this instance between INT and CRT_1/AKI). See: DAGitty. **UK1:** UnKnown/unacknowledged variable set 1; **SEX**: Sex; **DoB**: Date of birth-determined age at intervention; **DM**: Pre-intervention diagnosis of diabetes mellitus; **CRF**: Pre-intervention diagnosis of chronic renal failure; HCT_0: Baseline/pre-intervention haematocrit values; CRT_0: Baseline/pre-intervention creatinine values; **IP**: Indication for the procedure (venous vs. arterial); **UK2:** UnKnown/unacknowledged variable set 2; **INT**: Intervention (PPT vs. CDT); **SURG**: Post-intervention open surgical procedures; HCT_1: Post-intervention haematocrit values; CRT_1: Post-intervention creatinine values; **AKI**: Acute kidney injury; **UK3**: UnKnown/unacknowledged variable set 3

— albeit only when the principal aim of the DAG is to explore or address potential sources of analytical and inferential bias when estimating a *single* focal relationship:

- **Step 10** — reduce the number of alternative DAGs generated at Step 9 by creating three discrete covariate 'sets' comprising all those whose temporal position can confidently be located: before the specified exposure variable; after the specified outcome variable; or in-between the two.

To this end, Fig. 5 provides four further iterations of the DAG summarised in Fig. 3 in which each covariate has been included in one of three discrete sets of variables located: before the specified exposure variable (INT); after the specified outcome variable (CRT_1/AKI); or in-between the two. Not only can this approach substantially reduce the number of alternative DAGs required to accommodate uncertainty in the temporal positioning of covariates *within* each of these three periods; but it does so without recourse to aetiological knowledge, or professional/operational experience (and the risk of cognitive bias that reliance thereon can entail). In the case of Escobar et al.'s [25] study, where only two of the available covariates (SURG and HCT_1) were considered likely to have occurred in *more than one* of these three periods of time (see Sect. 2.3 and Figure 4, above), only four alternative DAGs are required to summarise the key causal relationships between each of the covariates and the specified exposure and outcome variables — a substantial simplification of the 240+ possible DAGs considered necessary following Step 9 (above).

3 Using DAGs to Identify/Address Potential Sources of Analytical or Inferential Bias

3.1 Identifying Potential Errors and Biases Associated with Epistemological Assumptions

A key benefit of using DAGs to represent disease and healthcare pathways (and the data generating mechanisms involved therein) is that these representations can help to expose potential errors and biases associated with the epistemological assumptions that all such diagrams require. Indeed, we have already seen how DAGs informed by temporal logic might help to assess whether the causal paths considered likely or necessary on the basis of prevailing aetiological knowledge, and professional/operational experience are consistent with those considered possible or plausible on the basis of temporality and temporal logic. These assessments of internal and external consistency can inform, or be prompted by, decisions made at each and every stage of drawing a DAG. For example, when the variables considered relevant to include are arranged in a temporal-causal sequence, it soon becomes apparent if *none* of the covariates act: as plausible, probabilistic causes of the exposure and/or outcome; or as variables that fall along the causal path between the two. The most helpful DAGs in this regard will have covariates located before, after *and* in-between the specified exposure and outcome variables in order to prompt reflection on the specific roles that variables positioned in each of these locations might perform (and any associated risks of conditioning-dependent bias when estimating the strength, direction and precision of the focal relationship involved — see Sect. 3.2, below). Moreover, the absence or omission of covariates within any of these locations should prompt substantial reflection and discussion amongst the healthcare students, practitioners and researchers concerned regarding what additional variables (whether measured or unmeasured, known or unknown) might be relevant, appropriate, helpful or necessary to include, including those that might as yet be unacknowledged (such as UK1, UK2 and UK3 in Figs. 3, 4 and 5).

3.2   Identifying Potential Sources of Conditioning-Dependent Analytical and Inferential Bias

From an analytical and statistical perspective, an even more important benefit of using DAGs to represent disease and healthcare pathways, and the underlying data generating mechanisms these pathways involve, is that they help to identify potential sources of conditioning-dependent bias that can affect the estimated strength, direction and precision of any specified focal relationship [41, 42]. These biases are the parametric consequences of the temporal-causal relationships that exist between each of the covariates and the specified exposure and outcome variables. Where these covariates precede both the specified exposure and specified outcome variables they will act as potential 'confounders' (i.e. plausible, probabilistic causes of *both* the exposure and the outcome [41]) and — unless their effects are removed by conditioning on the covariates concerned (whether through selective sampling, stratification or statistical adjustment [43]) — the estimated strength, direction and precision of the specified focal relationship may then be susceptible to 'confounding bias'. In contrast, wherever covariates are plausible, probabilistic *consequences* of both the exposure and the outcome (i.e. where they might be called 'consequences of the outcome' [16]), or where they act as 'mediators' that fall along the speculative causal path between the exposure and outcome (being plausible, probabilistic *consequences* of the exposure, and plausible, probabilistic *causes* of the outcome [44]), the so-called 'collider bias' that these can contribute to the estimated strength, direction and precision of the specified focal relationship will *only* occur following conditioning thereon (whether, as before, through selective sampling, stratification or statistical adjustment [45–48]).

DAGs therefore have substantial utility for: classifying which of the covariates might act as confounders, consequences of the outcome or mediators; and thereby evaluating the potential contribution that each might make to conditioning-dependent analytical and inferential bias when estimating a specified focal relationship. In particular, DAGs help to prompt careful reflection amongst the healthcare students, practitioners and researchers involved regarding: the potential risk of conditioning on consequences of the outcome or mediators as a result of sampling, stratification or statistical adjustment decisions (whether deliberately, unintentionally or as a result of misclassification [43]); and whether the inclusion, measurement and adjustment for alternative confounders (and particularly those offering more accurate and more comprehensive assessments of the potential for confounding) might help further mitigate the risk of residual or unadjusted confounding bias, respectively (see [17, 18]; and [16]).

For example, returning to the DAG presented in Fig. 3 (based on the DAG provided by Escobar et al. [25]: Figure 7: 244 PA; 29 AAM), the authors concerned might have found it insightful to consider whether:

- the estimated relationship between the specified exposure variable (INT) and the specified outcome variable (CRT_1/AKI) might have been subject to collider bias as a result of conditioning on one or more of: the measured mediators (i.e. SURG; and HCT_1); the set of unacknowledged mediators (i.e. one or more covariates included in UK2); or the set of unacknowledged consequences of the outcome (i.e. one or more of the covariates included in UK3);
- it might be worth conditioning on the additional potential confounders (SEX; CRT_0; and HTC_0) — that were measured yet were omitted from (and could have been included in) Escobar et al.'s [25] original DAG — in order to further mitigate the likely effects of unmeasured and unadjusted confounding bias [49];
- it might likewise be worth conditioning on any additional unmeasured potential confounders, including those as yet unacknowledged (i.e. those within UK1) — assuming either that: data on appropriate variables might be available within the medical records of the patients examined; or it would be feasible to repeat the analyses with a subsequent cohort of patients for whom some of these additional covariates might then be identified and measured; and
- any of the covariates situated immediately *before* and immediately *after* the specified exposure variable (such as: IP, CRT_0 and HTC_0; and SURG and HCT_1, respectively) might have *actually* occurred or crystallised later/earlier (i.e. *after* or *before* the specified exposure variable) — resulting in: conditioning on mediators that had been misclassified as potential confounders (in this instance: IP; CRT_0; and/or HTC_0); or a failure to condition on potential confounders that had been misclassified as mediators (in this instance: SURG; and/or HCT_1).

3.3 Mitigating Potential Sources of Bias When Selecting the Datasets and Statistical Adjustment Sets Used

In each of these instances, a DAG (such as that presented in Fig. 3) can clearly help provide a credible, principled and interrogable basis for exploring whether improvements in a number of key methodological decisions might help to mitigate the risk of analytical and inferential bias during the estimation of specified focal relationships — including:

- the sampling or stratification of participants included in a prospective study — to reduce the risk of collider bias as a result of sampling or stratification that deliberately or unintentionally conditions on a mediator or consequence of the outcome;

- the choice of any retrospective or existing dataset(s) selected for secondary analysis — again to reduce the risk of collider bias by carefully ensuring that any datasets selected for analysis are unaffected by sampling or stratification that might have deliberately or unintentionally conditioned on a mediator or a consequence of the outcome;
- the measurement of additional or alternative covariates (either prospectively or by those involved in generating the existing dataset[s] subsequently selected for secondary analysis) — to reduce the risk of inappropriate conditioning on mediators or consequences of the outcome misclassified as confounders; and to mitigate residual or unmeasured (and therefore unadjusted) confounding bias through statistical adjustment of more accurately measured or a more comprehensive selection of potential confounders, respectively; or
- the exclusion of misclassified confounders, and the inclusion of more accurately measured (or a more comprehensive selection of additional or alternative) confounders in the covariate adjustment sets of the multivariable statistical models used to estimate the strength, direction and precision of the specified focal relationship — again to reduce the risk of inappropriate conditioning on mediators or consequences of the outcome misclassified as confounders; and to mitigate residual or unmeasured (and therefore unadjusted) confounding bias through statistical adjustments of more accurately measured, or a more comprehensive selection of, potential confounders, respectively.

3.4   Identifying Potential Sources of Bias During the Critical Appraisal of Published Analyses

Each of these potential improvements in methodological decisions that DAGs are able to support also lend themselves to the critical appraisal of previously published analyses, regardless of whether these made use of DAGs (and did so competently or otherwise). Indeed, after extracting the variables that were included by Escobar et al. [25] in their DAG (Fig. 7: 244 PA; 29 AAM) and/or Table 1 (241 PA; 21 AAM), and after identifying the specified exposure (INT) and outcome (AKI) variables involved in their study's principal focal relationship, the worked example presented in this chapter illustrates how it can be relatively straightforward to: generate a DAG (even when the authors concerned did not use or report one, as Escobar et al. [25] did) using each of the ten steps described in Sect. 2, above; and classify which of the measured and/or included covariates might act as confounders (in this instance: DoB; Sex; DM; CRF; CRT_0; HCT_0; and IP), potential mediators or consequences of the outcome (in this instance: SURG; and HCT_1) for the focal relationship concerned (see Fig. 5). Based on this information, and after careful consideration of the data sources, sampling techniques, measurement protocols and primary/secondary datasets used, it should also be possible to assess whether any of the estimates generated for the principal focal relationship(s) examined might have been affected by prospective study designs or a reliance on existing, secondary datasets that involved:

- sampling or stratification (whether deliberate or unintentional) that was likely to have conditioned on a mediator or consequence of the outcome — thereby increasing the risk that the estimates generated might have been affected by collider bias;

- inadequate numbers of covariates, or covariates measured with insufficient accuracy or precision, to permit robust conditioning on/adjustment for those classified as confounders — thereby increasing the risk that the estimates generated were affected by unmeasured/unadjusted or residual confounding, respectively; and

- the misclassification of, and inappropriate conditioning on, mediators or consequences of the outcome as if these were confounders — thereby increasing the risk that the estimates generated were inadvertently affected by collider bias as a result of conditioning on these (misclassified) colliders.

4 A Role for Educators in Addressing the Limitations of Directed Acyclic Graphs

4.1 Transparent Reporting of Context-Specific Uncertainties, Assessments and Potential Errors

DAGs can only ever provide imperfect *theoretical* approximations of causal pathways and data generating mechanisms, not least because — except where these pathways and mechanisms are deliberate or incidental human artefacts that have been faithfully rendered or comprehensively described, respectively — it is simply implausible that we will ever know: *all* of the relevant variables involved; *which* of these are necessary or sufficient to include therein; and *where* each of these should be temporally and causally positioned with respect to one another. Indeed, beyond the challenges posed by omitted, time-variant, non-asynchronous and temporally obscure variables, it is also likely that many phenomena, characteristics and features (*both* time-invariant *and* time-variant) occur or crystallise over a sufficiently broad period of time that their causally relevant temporal positions within a DAG overlap with (many) other variables, making it very difficult to determine which (if any) are causes or consequences of each of the others. Wherever the extent of these overlaps makes it unclear whether the variables concerned act as confounders rather than mediators or consequences of the outcome, it will severely limit the practical utility of the DAGs concerned. This is because such DAGs will involve substantial uncertainty as to which of the covariates require conditioning (to mitigate confounding bias), and which do not (to avoid collider bias).

These epistemological uncertainties remind us that DAGs, like all tools, are rarely infallible. Moreover, as principled yet highly conceptual models, DAGs can prove elusive and difficult to articulate and operationalise; and the level of skill, diligence and discipline required to grasp and apply the principles on which DAGs are based can impose very real limitations on their practicability and the value of the evidence they generate. As such — and like any novel tool contingent on the knowledge, understanding and competence required for their successful application — DAGs can somewhat paradoxically: reduce (rather than enhance) the thoughtfulness involved in the design, application and interpretation of inferential statistics; obfuscate or conceal thoughtless and incompetent practices; offer an entirely false aura of novelty, sophistication and expertise; and make it *more* (not *less*) difficult for others to critique and challenge the theories and analyses involved, even when these are patently misguided or flawed [50]. Indeed, the recent review by Tennant et al. [35] found substantial evidence of simplistic, wrong-headed, thoughtless and ostensibly sloppy applications of DAGs, including: a remarkably large proportion of articles (38%) that reported using DAGs but failed to share them (thereby eliminating the possibility of evaluating the theoretical integrity or temporal plausibility of these DAGs); very few (<3%) that included all of the plausible, probabilistic causal paths; and fewer than half that reported either the DAG-implied covariate adjustment set(s) (49%) *or* the results of multivariable statistical models where these adjustment sets had been faithfully applied (43%).

These weaknesses in the application and reporting of DAGs mean that their growing popularity may not necessarily lead to consistent improvements in the elimination, mitigation or acknowledgement of bias resulting from inappropriate conditioning on: one or more colliders; or an inadequate selection of incorrectly classified, or imprecisely measured, confounders. Their use is also unlikely to *successfully* bridge the divide between professional expertise and statistical inference *without*: substantial improvements in practice; and greater clarity and transparency in reporting. To this end, Tennant et al. [35] offered eight succinct recommendations to improve both the application and reporting of DAGs; and offered a useful checklist to assist authors preparing (and reviewers evaluating) articles using DAGs (Supplementary Table S6). These have been summarised in the appendices to this chapter (see Appendix 2), and augmented with two further recommendations for improving the interpretation and interrogability of DAGs, namely: the importance of reporting the intended *application* of the DAG (whether analytical, speculative or hypothetical); and using accessible software (such as DAGitty [51–53]) when drawing and sharing DAGs, to avoid the limitations of static two-dimensional visual representations which can be difficult to read, and particularly so when they contain more than a modest number of variables and causal paths (compare, for example, Figs. 2, 3, 4 and 5).

4.2   A Role for Educators in Optimising the Utility of Directed Acyclic Graphs

The aim of this chapter has been to encourage the adoption, and improve the application, of DAGs by demonstrating how educators might use them to facilitate the acquisition of secondary/ancillary skills in inferential statistics by healthcare students, practitioners and researchers. The rationale for this approach stems from the dual utility of DAGs as principled; qualitative, nonparametric representations of temporal-causal disease and healthcare pathways; and of the underlying data generating mechanisms these involve — i.e. the mechanisms on which the quantitative, parametric insights generated using inferential statistics depend. While DAGs require thoughtfulness and a willingness to accommodate uncertainty, they help to prompt careful reflection and offer far greater transparency regarding the assumptions involved when applying inferential statistics to the analysis of essentially theoretical (and often extensively speculative or entirely hypothetical) disease and healthcare pathways.

These benefits accrue because DAGs have substantial utility for bridging the divide between: the aetiological knowledge and professional/operational experience on which contemporary clinical training and practice depend; and the more abstract epistemological and analytical considerations required to extract robust statistical insight from health and healthcare data. Moreover, provided it is plausible to generate finite and manageable numbers of alternative DAGs that accommodate any uncertainty regarding the temporal positioning of critical covariates (particularly those that might plausibly occur before or after the specified exposure variable), DAGs can still offer a coherent approach to the assessment and application of alternative conditioning strategies involving sampling, stratification or adjustment; and thereby support bias-mitigated estimation of the strength, direction and precision of the focal relationship(s) examined. Indeed, where these estimates are largely unaffected by the speculative re-positioning of covariates that might plausibly occur or crystallise before vs. after the specified exposure variable (as might occur where the impact of the confounder and collider bias involved is weak), then the use of multiple, alternative DAGs may even provide substantial reassurance that the estimates generated are likely to be robust, and therefore have substantial practical utility.

For these reasons, DAGs represent an invaluable educational tool for helping to improve both the theoretical and analytical capabilities of healthcare students, practitioners and researchers by:

- obliging them to carefully specify any explicit and implicit assumptions that underpin the theoretical causal pathways on which their analytical designs are based;
- enabling them to explore the temporal plausibility, and any associated uncertainty, regarding the role(s) that measured, unmeasured, unmeasurable and unacknowledged covariates might play therein;
- helping them to address any temporal implausibility and uncertainty in the sampling, measurement and analytical assessments and decisions they make; and
- ensuring they can then acknowledge any residual risk of conditioning-dependent analytical and inferential bias when interpreting and reporting their findings.

The step change in capability, transparency and epistemological competency that each of these benefits provide should offer a compelling case for educators to integrate DAGs into the statistical training and support provided to healthcare students, practitioners and researchers. To facilitate this, the worked example summarised in this chapter can be used to generate a lesson plan or self-directed learning exercise covering each of the ten steps for drawing a DAG using temporal logic (as outlined in Sect. 2, above), and to extend the application of each of these steps beyond the critical appraisal of published analyses that focus on the estimation of (one or more) focal relationships, to demonstrate how related considerations might be usefully applied to: inform prospective studies relying on either secondary or primary datasets; and ensure these datasets are selected or generated so as to optimise the opportunities available for eliminating or mitigating analytical and inferential bias when estimating a clearly specified focal relationship.

Acknowledgements Johannes Textor and Mark Gilthorpe kindly shared many of the perspectives and insights (and much of the understanding) required to develop the approach for drawing DAGs presented in this chapter. I would also like to thank Peter Tennant and Laust Mortensen for sharing the approaches they use when explaining the meaning and utility of DAGs; and Tomasina Stacey, Renata Medeiros Mirra and Damian Farnell for their helpful comments and suggestions on previous versions of the chapter.

Appendix 1: A Comprehensive List of Variables Considered Relevant by Escobar et al. [25]

Although the worked example summarised in this chapter focussed only on those variables that Escobar et al. [25] chose to include in their published DAG (Figure 7: 244 PA; or 29 AAM) and Table 1 (241 PA; or 21 AAM), this constitutes a gross simplification of Step 1 ('Compile a comprehensive list of any phenomena, characteristics or features [and their associated variables] considered relevant to the focal relationship.'). Indeed, the simplified approach to Step 1 adopted in this chapter is only likely to be sufficient for critically appraising the temporal-causal plausibility of Escobar et al.'s [25] published DAG itself; and is unlikely to support a substantive assessment of any potential, additional and residual risks of analytical and inferential bias associated with the omission of relevant and important variables from their DAG. This is because whenever a more *comprehensive* assessment is required — of the variables and causal paths included and omitted from DAGs developed to support the estimation of specified focal relationships — it is necessary to generate a more *comprehensive* list of the phenomena, characteristics and features involved (*both* when designing DAG-informed studies/analyses *and* when evaluating their published findings).

In both instances, substantial reflection and discussion amongst the healthcare students, practitioners and researchers concerned can help to carefully explore all of the (known and unknown) phenomena, characteristics and features that might plausibly be relevant to the specified focal relationship(s) involved — and therefore what associated variables ought to be/have been included in the causal path diagram(s)/ DAG(s) used, whether to: generate (or select) primary (or secondary) datasets; or mitigate the risk of bias associated with the sampling or analytical constraints these datasets impose/involve. Substantial reflection and discussion may also be evident in the deliberations and decisions reported in published accounts of such studies, and these can provide a helpful list of any ancillary variables the authors considered relevant to each of the focal relationships and sampling frames involved (and, therefore, any additional variables that might have been useful or necessary to include in each of the studies' DAGs). Moreover — whenever the reviewer(s) involved in critically appraising such studies lacks the knowledge, insight and experience required to generate an extensive (or comprehensive) list of the phenomena, characteristics, features and associated variables likely to be relevant to/implicated in the focal relationship(s) concerned; or is particularly interested in the ideas, insights, views and perspectives provided by the authors involved — collating a detailed list of the variables cited in the study's rationale, design and interpretation (such as those summarised in the article's Introduction and Discussion sections) can prove invaluable.

To illustrate what this can entail when critically appraising published studies using DAG-informed designs and analyses, it is worth compiling a list of all of the phenomena, characteristics, features and associated variables that Escobar et al. [25] cited when developing, implementing, interpreting and reporting their study. Excluding those required to record the specified exposure and outcome variables, this list of variables includes any mentioned as potentially relevant for consideration in the study's rationale, design and interpretation (such as those summarised in the article's Introduction or Discussion sections), as well as any variables that were *actually* available to, and were measured, summarised and/or analysed in the Methods and Results sections of the article.

Careful and close reading of Escobar et al.'s [25] article identified numerous phenomena, characteristics, features and variables that had been cited as relevant to their principal focal relationship. These included those considered likely to: reflect explicit or implicit inclusion and exclusion criteria (on the basis of which the patients examined were selected for analysis); determine whether the patients concerned received each of the two clinical procedures/interventions (PPT vs. CDT); mediate the impact of these clinical interventions on the specified outcome (AKI); and emerge as immediate or longer term sequelae of the clinical procedure/intervention received (and/or the specified outcome, AKI itself).

Excluding the specified exposure and outcome variables, covariates relevant to each of these phenomena, characteristics and features were cited in: every section of the article (including the Introduction, Methods, Results and Discussion sections); both Tables; and all seven Figures, as summarised below:

Introduction: 'acute arterial thrombotic syndrome'; 'acute venous thrombotic syndrome'; 'treatment time [duration]'; 'total volume of fibrinolytics'; 'intensive care unit length of stay'; 'hospital length of stay'; 'hospital costs'; 'bleeding complications'; 'hemolysis'; 'hematuria'; 'renal failure'; 'renal function'; 'dialysis'.

Methods: 'patients at our institution'; 'from 2007 to 2013 [hence date on which treatment was provided]'; 'The approach of treatment was up to the practitioner at the time of the [procedure/]intervention, and not protocoled [hence attending physician/practitioner]'; 'demographics, indications, laboratory values before and after the procedure[/intervention] (up to 3 days) [see below for the specific variables involved]'; 'chronic kidney disease'; 'any patient with that diagnosis documented by a nephrologist before our procedure[/intervention — hence documenting nephrologist]'; 'baseline creatinine (Cr)'; 'contrast-induced nephropathy (CIN)'; 'increase in Cr [creatinine] within 72 hours [of the procedure/intervention]'; 'patients on dialysis before AJ'; '[patients with] duplicated codes'; '[patients] without laboratory values obtained before and 24–72 hours after treatment[/procedure/intervention]'; 'hematocrit (HCT)'; 'glomerular filtration rate (GFR)'; '[patient] age'; '[patient] weight'; 'expos[ure] to 270 mgI/mL Iodixanol … a hypo-osmolar, non-ionic iodinated contrast agent'.

Results: 'Cr kinase'; 'myoglobin'; '[patient] age'; 'baseline Cr'; 'HCT before treatment'; 'arterial thrombolysis … indication [for the procedure/intervention]'; 'venous [thrombolysis … indication for the procedure/intervention]'; 'rise in Cr'; 'preoperative chronic renal failure (CRF)'; 'dialysis within 2 days of … procedure/intervention'; 'died [mortality] during hospitalization'; 'mesenteric ischemia'; 'pulmonary embolus'; 'open surgery'; 'thromboembolectomy'; 'fasciotomy[y]'; 'bypass or endarterectomy'; 'major amputation'; '[patient sex] male [vs. female]'; 'diabetes mellitus (DM)'; 'postprocedure[/intervention] HCT'; 'blood loss [as a synonym for change in HCT?] … percentage change from baseline'.

Table I: '[patient] age'; 'male [patient sex]'; 'diabetes'; '[baseline] Cr'; 'arterial indication [for procedure/intervention]'; 'venous indication [for procedure/intervention]'; '[baseline] HCT'.

Table II: 'angioplasty [endovascular procedure/intervention in addition to AJ or CDT]'; 'stenting [endovascular procedure/intervention in addition to AJ or CDT]'; 'fasciotomy [post-procedure/intervention open surgical procedure/intervention]'; 'thromboembolectomy [post-procedure/intervention open surgical procedure/intervention]'; 'bypass/endarterectomy [post-procedure/intervention open surgical procedure/intervention]'; 'major amputation [post-procedure/intervention open surgical procedure/intervention]'.

Figure 1: 'hematuria'.

Figure 2: 'Cr [change from baseline post-procedure/intervention]'.

Figure 3: 'change in creatinine [from baseline post-procedure/intervention]'; 'open surgery [post-procedure/intervention]'.

Figure 4: 'change in hematocrit [from baseline post-procedure/intervention]'.

Figure 5: 'maj surg … major surgery [major open surgery procedures/interventions]'; '[patient] age'; 'male … [patient sex]'; 'diabetes'; 'CRF … chronic renal failure'; 'drop in HCT … hematocrit [from baseline post-procedure/intervention]'.

Figure 6: 'maj surg … major surgery [major open surgery procedures/interventions]'; '[patient] age'; 'male … [patient sex]'; 'diabetes'; 'CRF … chronic renal failure'; 'drop in HCT … hematocrit [from baseline post-procedure/intervention]'.

Figure 7: 'venous clot [indication for procedure/intervention]'; 'arterial clot [indication for procedure/intervention]'; 'CRF [chronic renal failure]'; '[patient] age'; 'DM [diabetes mellitus]'; 'drop_hematocrit [from baseline post-procedure/intervention]'; '[open] surgery [post-procedure/intervention]'.

Discussion: 'as young as … [patient age]'; 'diabetes'; 'final HCT [post-procedure/intervention]'; 'change in HCT … drop in HCT [from baseline]'; '[open] surgical procedures[/interventions]'; 'anemia'; 'transfusion'; 'iatrogenic [hemo]dilution'; 'Cr … fallen … change [from baseline]'; 'hemolysis'; 'renal failure'; 'free serum haemoglobin'; 'plasma haptoglobin'; 'heme'; 'globin'; 'Tamm-Horsfall proteins'; 'intra[renal]tubular casts'; '[renal] tubular flow'; 'oliguria'; 'azotemia'; 'venous occlusions'; 'renal function'; 'thromboembolic complications

Table 1 Variables cited by Escobar et al. [25] with reference to study rationale, design and/or analysis

Phenomena, characteristics, features and associated variables mentioned/cited	Introduction	Methods	Results	Discussion	Table I	Table II	Figure 1	Figure 2	Figure 3	Figure 4	Figure 5	Figure 6	Figure 7
Acute arterial thrombotic syndrome	X		X	X	X								X
Acute venous thrombotic syndrome	X		X	X	X								X
Treatment duration	X			X									
Total volume of fibrinolytics	X												
Length of stay in intensive care unit	X												
Length of stay in hospital	X												
Hospital costs	X												
Bleeding complications	X												
Haemolysis	X			X									
Haematuria	X						X						
Renal (dys)function	X			X									
Renal failure	X			X									
Dialysis post-procedure	X		X	X									
Patients treated 'at our institution'		X											
Date of treatment (2007–2013)		X											
Treatment approach of attending practitioner		X											
Chronic kidney disease		X											
Documenting nephrologist		X											

Creatinine (Cr) at baseline	X	X		X								
Contrast-induced nephropathy (CIN)	X		X									
Change in Cr within 72 hours of procedure	X	X	X				X	X				
Dialysis prior to procedure	X											
Patients with duplicated codes	X											
Patients with missing laboratory values	X											
Haematocrit (HCT) at baseline	X	X		X								
Glomerular filtration rate (GFR)	X											
Age of patient	X	X	X	X						X	X	X
Weight of patient	X											
Exposure to 270 mgI/mL Iodixanol	X		X									
Cr kinase at baseline		X	X									
Cr kinase post-procedure		X	X									
Myoglobin at baseline		X	X									
Myoglobin post-procedure		X	X									
Chronic renal failure at baseline		X								X	X	X
Mortality during hospitalisation		X	X									
Mesenteric ischaemia		X										
Pulmonary embolus		X										

(continued)

Table 1 (continued)

Phenomena, characteristics, features and associated variables mentioned/cited	Introduction	Methods	Results	Discussion	Table I	Table II	Figure 1	Figure 2	Figure 3	Figure 4	Figure 5	Figure 6	Figure 7
Open surgery post-procedure (generic)			X	X					X		X	X	X
Thromboembolectomy			X			X							
Fasciotomy			X	X		X							
Bypass or endarterectomy			X			X							
Major amputation			X	X		X							
Sex of patient			X		X						X	X	
Diabetes mellitus (DM) at baseline			X	X	X						X	X	X
Change in HCT within 72 hours of procedure			X	X						X	X	X	X
Angioplasty				X		X							
Stenting						X							
HCT post-procedure				X									
Anaemia				X									
Transfusion				X									
Iatrogenic haemodilution				X									
Free serum haemoglobin				X									
Plasma haptoglobin				X									
Heme				X									
Globin				X									
Tamm-Horsfall proteins				X									
Intratubular casts				X									
Renal tubular flow				X									
Oliguria				X									
Azotaemia				X									

Venous occlusions			X							
Thromboembolic complications post-angioplasty			X							
Haemoglobinuria			X							
Limb ischaemia			X							
Dialysis dependence			X							
Acute renal tubular necrosis			X							
Renal hemosiderin deposits			X							
Intracranial haemorrhage			X							
Arterial blood flow			X							
Number of contrast boluses			X							
Rhabdomyolysis from ischaemia/reperfusion			X							
Recurrent thrombotic syndromes			X							
Time intervals between repeat procedures			X							
Duration of haematuria post-procedure			X							
Blood loss post-procedure			X							

after angioplasty'; 'hemoglobinuria'; 'renal dysfunction'; 'patients … who died [mortality]'; 'limb ischemia'; '[renal failure requiring] dialysis'; 'dialysis dependence'; 'acute [renal] tubular necrosis'; 'overwhelming intravascular hemolysis'; 'fulminant renal failure'; 'hemosiderin deposit[s] … in the kidneys'; 'intracranial hemorrhage'; 'arterial blood flow'; 'CIN [contrast-induced nephropathy]'; '270 mgI/mL Iodixanol'; 'time [duration] of treatment'; 'number of contrast boluses'; 'rhabdomyolysis from ischemia/reperfusion'; 'myoglobin [at baseline]'; 'myoglobin [post-procedure/intervention]'; 'Cr kinase [at baseline]'; 'Cr kinase [post-procedure/intervention]'; 'amputations'; 'fasciotomies'; 'venous thrombolysis [indication for procedure/intervention]'; 'arterial [thrombolysis indication for procedure/intervention]'; 'recurrent thrombotic syndromes'; 'time intervals between treatments [/procedures/interventions]'; 'duration of hematuria [post-procedure/intervention]'; 'blood loss'.

In total, perhaps as many as 75 discrete sampling criteria and covariates were cited within Escobar et al. ([25]; see Table 1), although a good number of these: may have been functionally synonymous (such as: 'renal failure' and 'renal (dys)function'); comprised categories or components of other covariates (such as: 'fasciotomy', 'thromboembolectomy', 'bypass/endarterectomy' and 'major amputation', which all appeared to have been classified as 'open surgical procedures'); or constituted derived variables based on repeated measures of one or more other covariates (such as: 'change in creatinine' or 'drop in HCT'). As such, the total number of separate sampling criteria and covariates is likely to be somewhat lower than the 75 listed in Table 1. Nonetheless, it is clear that a sizeable proportion of these (45/75; 60.0%) were only mentioned in just one section of the published article, and the number cited in the Methods and/or Results sections (35/75; 46.7%) may offer a fairer representation of the known sampling criteria and covariates for which measurements were actually available.

Appendix 2: Ten Recommendations to Improve the Application and Reporting of DAGs

The recommendations summarised below include eight proposed by Tennant et al. ([35]: 628–629; Supplementary Table S6) and two additional recommendations intended to improve the interpretation, presentation and interrogability of DAGs. The first eight of these recommend that:

1. *The focal relationship(s) examined (i.e. the relationship[s] between the specified exposure and outcome variables) should always be reported* — a critical consideration given the specification of the focal relationship(s) determines which of the covariates considered relevant for inclusion in the DAG are likely to operate as potential confounders, mediators or consequences of the outcome.
2. *The DAG(s) generated to inform the estimation of each focal relationship should always be reported* — again, a critical consideration given that these DAGs will indicate which of the covariates considered relevant for inclusion therein were/

should have been interpreted as operating as potential confounders, mediators or consequences of the outcome.

3. *Each of the DAGs provided should include all variables considered relevant to the causal pathway(s) and data generating mechanism(s) concerned, including those covariates where no measurements were available (and those that might otherwise remain unacknowledged) — which will ensure that the authors have* explicitly acknowledged the risk that any such covariates might have played as potential sources of collider bias during sampling or stratification, or as sources of unmeasured and unadjusted confounding.

4. *Every DAG should contain all of the plausible, probabilistic causal paths between each of the successive variables considered relevant to include therein, except where there is definitive evidence that any such path(s) cannot or do not exist (in which instance, such evidence should then be reported) — so that the* strong assumption invoked by the omission of (possible and plausible) probabilistic paths is not made without substantive justification; and it is possible to assess any rationale for omitting ostensibly plausible, probabilistic causal paths.

5. *The temporal-causal sequence of measured (and of unmeasured and unacknowledged) variables considered relevant to include in each DAG is best presented in a consistent or coherent direction (whether horizontally, vertically or diagonally) — so that it is easier to ensure that the DAG is both (uni)directional and* (strictly) *acyclic*.

6. *The DAG-implied covariate adjustment sets of any multivariable statistical models used to estimate the strength, direction and precision of specified focal relationships should be reported — so that it is possible to ensure that these* adjustment sets include: all measured covariates acting as potential confounders within the DAG; and exclude any covariates acting as mediators or consequences of the outcome, while transparently reporting and acknowledging any risk of unmeasured and unadjusted confounding from unmeasured and unacknowledged covariates acting as potential confounders.

7. *The estimated strength, direction and precision of each focal relationship generated using the precise covariate adjustment set as implied by the relevant DAG should be reported (or, at least, the most complete set available that contains all of the measured covariates likely to act as potential confounders) — to ensure* the insight generated by using a DAG to inform such analyses is reported for consideration and assessment.

8. *The use of alternative, modified or augmented covariate adjustment sets be explicitly justified, such as those containing: only a subset of the measured covariates acting as potential confounders within the DAG; or additional covariates omitted from the DAG, or not considered to be acting as potential confounders therein — thereby ensuring that the estimates produced when using such* adjustment sets can be appropriately interpreted and acknowledged as *potentially* biased.

While the first five of Tennant et al.'s [35] recommendations should all help health-care students, practitioners and researchers to draw, report and share their DAGs in a consistent, comprehensive and effective fashion, the last three will only be relevant to those studies where DAGs have been used to represent real-world causal processes and thereby inform the selection of covariate adjustment sets to mitigate the risk of confounding bias when estimating the strength, direction and precision of a specified focal relationship using multivariable statistical analyses. Indeed, there are two additional recommendations that might also have substantial relevance: to those who are unaware of the alternative applications of DAGs (i.e. as representations of entirely speculative or hypothetical pathways/mechanisms as opposed to analytical tools used to support inferential statistics [54]); or within analytical contexts and applications where the DAGs involved contain more than a handful of variables, and therefore suffer a substantial decline in readability when generated or shared simply as static, two-dimensional diagrams (see for example: Fig. 3; and [16]).

9. *Since DAGs can be developed to support analytical, speculative or hypothetical applications, students, practitioners and researchers using DAGs should report the intended application(s) for which their DAGs were designed, and the rationale involved when constructing these — this is* because the intended application(s) and the rationale involved will be central to evaluating the likely value, insight and inference that might be drawn from the design of studies and statistical models based on such DAGs.

10. *Any DAGs in which the number of variables and causal paths included make it unclear whether any of these are present or missing should be shared in interrogable formats that make it possible to ascertain the presence or absence of covariates and impossible vs. plausible/probabilistic causal paths therein —* since, in the absence of such formats, the DAG will provide neither an accurate nor a definitive summary of the variables and causal paths the authors intended to include. Examples of suitable formats include those generated using: WinBUGS [55]; Gephi [56]; or DAGitty [52].

These two recommendations have been applied to the worked example presented in this chapter. First, by making it clear that the DAGs drawn in Figs. 2, 3 and 5 were intended to represent real-world causal processes — i.e. the disease and healthcare pathways relevant to the diagnostic and treatment-related decisions examined by Escobar et al. [25] — and thereby support estimation of the focal relationship between INT and AKI. Second, by providing interrogable versions of these three DAGs redrawn using DAGitty software and published online therein (see: Fig. 2: http://dagitty.net/mp8lf8Z; Fig. 3: http://dagitty.net/mXS3gft); Fig. 5: http://dagitty.net/mjHTJJP — all of which were drawn, and are best viewed, in the 'SEM-like' option available under DAGitty's 'Diagram Style' tab).

References

1. Milic NM, Masic S, Milin-Lazovic J, Trajkovic G, Bukumiric Z, Savic M, Milic NV, Cirkovic A, Gajic M, Kostic M, Ilic A, Stanisavljevic D. The importance of medical students' attitudes regarding cognitive competence for teaching applied statistics: multi-site study and meta-analysis. PLoS One. 2016;11:e0164439, (1–13).
2. Hibbard JH, Peters E. Supporting informed consumer health care decisions: data presentation approaches that facilitate the use of information in choice. Annu Rev Public Health. 2003;24:413–33.
3. Sinatra GM, Kienhues D, Hofer BK. Addressing challenges to public understanding of science: epistemic cognition, motivated reasoning, and conceptual change. Educ Psychol. 2014;49:123–38.
4. Morris RL. Increasing specialization: why we need to make mathematics more accessible. Soc Epistemol. 2021;35:37–47.
5. Grimshaw JM, Eccles MP, Lavis JN, Hill SJ, Squires JE. Knowledge translation of research findings. Implement Sci. 2012;7:1–7.
6. Watt H. Statistics education beyond "significance": novel plain English interpretations to deepen understanding of statistics and to steer away from misinterpretations. In: Medeiros Mirra RJ, Farnell D, editors. Teaching biostatistics in medicine and allied health sciences. Berlin: Springer; 2022.
7. Chapman SJ, Grossman RC, FitzPatrick ME, Brady RR. Randomized controlled trial of plain English and visual abstracts for disseminating surgical research via social media. J Br Surg. 2019;106:1611–6.
8. Larson D, Mazur K, White D, Yarnall C. The user's guide project: looking back and looking forward. J Human Math. 2020;10:411–30.
9. Bartlett G, Gagnon J. Physicians and knowledge translation of statistics: mind the gap. Can Med Assoc J. 2016;188:11–2.
10. Kennedy H, Hill RL, Aiello G, Allen W. The work that visualisation conventions do. Inf Commun Soc. 2016;19:715–35.
11. Passera S. Flowcharts, swimlanes, and timelines: alternatives to prose in communicating legal–bureaucratic instructions to civil servants. J Bus Tech Commun. 2018;32:229–72.
12. Carnap R. Theory and prediction in science. Science. 1946;104:520–1.
13. Tromovitch P. The lay public's misinterpretation of the meaning of 'significant': a call for simple yet significant changes in scientific reporting. J Res Pract. 2015;11:1–11.
14. Castell S, Cameron D, Ginnis S, Gottfried G, Maguire K. Public views of machine learning. London: Ipsos MORI Social Research Institute, Ipsos MORI; 2017. p. 1–92.
15. Hanel PH, Mehler DM. Beyond reporting statistical significance: identifying informative effect sizes to improve scientific communication. Public Underst Sci. 2019;28:468–85.
16. Ellison GTH. Might temporal logic improve the specification of directed acyclic graphs (DAGs)? J Stat Data Sci Educ. 2021;29:202–13.
17. Ellison GTH, Harrison W, Law GR, Textor J. Graphical, cross-tabulatory and relational: which is best for drawing DAGs? Proc VI Eur Congr Methodol. 2014a;6:11.
18. Ellison GTH, Harrison W, Law GR, Textor J. Teaching DAGs to support MBChB students design, analyze and critically appraise clinical research. Proc 35th Annu Burwalls Meet Teach Med Stat. 2014b;35:3.
19. Ben-Shlomo Y, Fallon U, Sterne J, Brookes S. Do medical students with A-level mathematics have a better understanding of the principles behind evidence-based medicine? Med Teach. 2004;26:731–3.
20. Murray S, Gal I. Preparing for diversity in statistics literacy: institutional and educational implications. Proc Sixth Int Conf Teach Stat. 2002;6:1–8.
21. Thompson R, Wylie J, Mulhern G, Hanna D. Predictors of numeracy performance in undergraduate psychology, nursing and medical students. Learn Individ Differ. 2015;43:132–9.

22. Martens M, Ammar A, Riutta A, Waagmeester A, Slenter DN, Hanspers K, Miller RA, Digles D, Lopes EN, Ehrhart F, Dupuis LJ, Winckers LA, Coort SL, Willighagen EL, Evelo CT, Pico AR, Kutmon M. WikiPathways: connecting communities. Nucleic Acids Res. 2021;49:D613–21.

23. Manktelow M, Iftikhar A, Bucholc M, McCann M, O'Kane M. Clinical and operational insights from data-driven care pathway mapping: a systematic review. BMC Med Inform Decis Mak. 2022;22:1–22.

24. Young T, Morton A, Soorapanth S. Systems, design and value-for-money in the NHS: mission impossible? Fut Healthc J. 2018;5:156–9.

25. Escobar GA, Burks D, Abate MR, Faramawi MF, Ali AT, Lyons LC, Moursi MM, Smeds MR. Risk of acute kidney injury after percutaneous pharmacomechanical thrombectomy using AngioJet in venous and arterial thrombosis. Ann Vasc Surg. 2017;42:238–45. See: published article [PA]; and author accepted manuscript [AAM].

26. Ensmenger N. The multiple meanings of a flowchart. Inform Cult. 2016;51:321–51.

27. Law GR, Green R, Ellison GTH. Confounding and causal path diagrams. Chapter 1. In: Tu YK, Greenwood DG, editors. Modern methods for epidemiology. Dordrecht: Springer; 2012. p. 1–13.

28. Althubaiti A. Information bias in health research: definition, pitfalls, and adjustment methods. J Multidiscip Healthc. 2016;9:211–7.

29. MacLure M, Schneeweiss S. Causation of bias: the episcope. Epidemiology. 2001;12:114–22.

30. Etchells E. Anchoring bias with critical implications. AHRQ Morbidity and Mortality Rounds on the Web; 2015, p 1–26. https://psnet.ahrq.gov/sites/default/files/import/webmm.ahrq.gov.350_slideshow.ppt.

31. Rehana RW, Huda N. A common heuristic in medicine: anchoring. Ann Med Health Sci Res. 2021;11:1461–3.

32. Ly DP. The influence of the availability heuristic on physicians in the emergency department. Ann Emerg Med. 2021;78:650–7.

33. Redelmeier DA, Ng K. Approach to making the availability heuristic less available. Br Med J Qual Saf. 2020;29:528–30.

34. Aneshensel CS. Theory-based data analysis for the social sciences. London: SAGE Publications; 2002.

35. Tennant PWG, Murray EJ, Arnold KF, Berrie L, Fox MP, Gadd SC, Harrison WJ, Keeble C, Ranker LR, Textor J, Tomova GD, Gilthorpe MS, Ellison GTH. Use of directed acyclic graphs (DAGs) to identify confounders in applied health research: review and recommendations. Int J Epidemiol. 2021;50:620–32.

36. Geer D. Unknowable unknowns. IEEE Secur Priv. 2019;17:80–79.

37. Molony DA. Cognitive bias and the creation and translation of evidence into clinical practice. Adv Chronic Kidney Dis. 2016;23:346–50.

38. Ellison GTH, De Wet T. Johannesburg's 'poor housing, good health' paradox: the role of health status assessment, statistical modelling, residential context and migrant status. Public Health. 2020;186:257–64.

39. Ellison GTH, Mattes RB, Rhoma H, De Wet T. Economic vulnerability and poor service delivery made it more difficult for shack-dwellers to comply with COVID-19 restrictions. S Afr J Sci. 2022;118:1–5.

40. Fleischer NL, Roux AD. Using directed acyclic graphs to guide analyses of neighbourhood health effects: an introduction. J Epidemiol Community Health. 2008;62:842–6.

41. Tennant PWG, Arnold K, Berrie L, Ellison GTH, Gilthorpe MS. Advanced modelling strategies: challenges and pitfalls in robust causal inference with observational data. Leeds: Leeds Institute for Data Analytics (LIDA); 2017. ISBN: 978-1-5272-1208-4.

42. van Zwieten A, Tennant PW, Kelly-Irving M, Blyth FM, Teixeira-Pinto A, Khalatbari-Soltani S. Avoiding overadjustment bias in social epidemiology through appropriate covariate selection: a primer. J Clin Epidemiol. 2022;149:127–36.

43. Blair A, Stewart P, Lubin JH, Forastiere F. Methodological issues regarding confounding and exposure misclassification in epidemiological studies of occupational exposures. Am J Ind Med. 2007;50:199–207.
44. Richiardi L, Bellocco R, Zugna D. Mediation analysis in epidemiology: methods, interpretation and bias. Int J Epidemiol. 2013;42:1511–9.
45. Cole SR, Platt RW, Schisterman EF, Chu H, Westreich D, Richardson D, Poole C. Illustrating bias due to conditioning on a collider. Int J Epidemiol. 2010;2010(39):417–20.
46. Elwert F, Winship C. Endogenous selection bias: the problem of conditioning on a collider variable. Annu Rev Sociol. 2014;40:31.
47. Munafò MR, Tilling K, Taylor AE, Evans DM, Davey Smith G. Collider scope: when selection bias can substantially influence observed associations. Int J Epidemiol. 2018;47:226–35.
48. Schisterman EF, Cole SR, Platt RW. Overadjustment bias and unnecessary adjustment in epidemiologic studies. Epidemiology. 2009;20:488–95.
49. VanderWeele TJ, Arah OA. Bias formulas for sensitivity analysis of unmeasured confounding for general outcomes, treatments, and confounders. Epidemiology. 2011;22:42–52.
50. Krieger N, Davey Smith G. The tale wagged by the DAG: broadening the scope of causal inference and explanation for epidemiology. Int J Epidemiol. 2016;45:1787–808.
51. Ankan A, Wortel IM, Textor J. Testing graphical causal models using the R package "dagitty". Curr Protoc. 2021;1:e45. https://doi.org/10.1002/cpz1.45, (1–22).
52. Textor J, Hardt J, Knüppel S. DAGitty: a graphical tool for analyzing causal diagrams. Epidemiology. 2011;22:745.
53. Textor J, van der Zander B, Gilthorpe MS, Liśkiewicz M, Ellison GTH. Robust causal inference using directed acyclic graphs: the R package 'dagitty'. Int J Epidemiol. 2016;45:1887–94.
54. Ellison GTH. The strengths and weaknesses of directed acyclic graphs (DAGs) as cognitive, analytical and educational tools for medical statistics. preprints.org. 2022;2 Sep:1–17. https://doi.org/10.20944/preprints202210.0084.v1.
55. Lunn DJ, Thomas A, Best N, Spiegelhalter D. WinBUGS-a Bayesian modelling framework: concepts, structure, and extensibility. Stat Comput. 2000;10:325–37.
56. Bastian M, Heymann S, Jacomy M. Gephi: an open source software for exploring and manipulating networks. Proc Int AAAI Conf Web Soc Media. 2009;3:361–2.

Statistics Without Maths: Using Random Sampling to Teach Hypothesis Testing

Owen Bodger

1 Introduction

1.1 Statistics and Anxiety

Anyone with experience in teaching statistics is aware of the negative impression many students have of the subject. This is well documented in the literature, with Onwuegbuzie and Wilson [1] commenting that it is often seen as "anxiety-inducing" and Dunn et al. [2] going as far as to suggest that some students viewed the courses as "something to be endured" and that they may "prefer to watch paint dry". Some have even claimed that this anxiety is one of the major barriers faced by teachers of the subject [3] with others suggesting that it might affect as many as 80% of university students [1].

Many reasons have been suggested for this, a summary of which can be found in Bromage et al. [4]. Levpušček and Cukon [5] argue that the attitude of a student towards mathematics is most important. They go further and suggest that interventions to address statistics anxiety should focus on their negative beliefs about their mathematical skills. Hulsizer and Woolf [6] make a similar observation and argue that maths anxiety should be taken into account when designing courses.

Regardless of the causes or manifestation of these negative feelings, we need to confront the reality of these feelings as teachers of the subject to non-specialists. This is a challenge made greater by the fact that few of our students have explicitly chosen to study our subject.

O. Bodger (✉)
Health Data Science, Institute of Life Science, Swansea University, Swansea, UK
e-mail: o.bodger@swansea.ac.uk

D. J. J. Farnell, R. Medeiros Mirra (eds.), *Teaching Biostatistics in Medicine and Allied Health Sciences*, https://doi.org/10.1007/978-3-031-26010-0_7

1.2 Recommendations on Teaching Statistics to Non-specialists

While many statistics teachers report finding guidelines difficult, some help and advice can be found. The American Statistical Association (ASA) have endorsed several reports, including the Guidelines for Assessment and Instruction in Statistics Education (GAISE) reports [7–9], which were intended to develop comprehensive guidelines on the teaching of statistics.

The first report [7] focussed on the teaching of statistics in schools, while the second [8] is probably of most interest to us as it addresses the teaching of introductory statistics in college. They recognise the variation in student audiences, but they argue that their recommendations apply even beyond the introductory level.

Their recommendations include teaching statistical thinking, integrating real data, incorporating technology and using assessments to improve learning. Their recommendations to focus on conceptual understanding and to foster active learning are of most relevance to this chapter though.

They were very clear that the students' depth of understanding of key concepts was more important at the expense of the breadth of their knowledge of methods. Furthermore, they stressed that teachers should utilise technology to automate calculations to allow greater emphasis on understanding. Later, they discuss topics that may be safely avoided, and these include manually performing the calculations behind hypothesis tests, arguing that these are "no longer necessary" and do not "reflect modern practice".

The second key component that is quite relevant to our teaching was the focus on active learning, described as involving students in active exploration of the subject. Among the suggestions they made to teachers is the use of physical exploration including the use of dice rolling or the drawing of cards.

2 Our Teaching Approach

2.1 Statistics at Swansea Medical School

Despite being located in a medical school, we do not carry out much direct teaching to the medical students. We teach introductory statistics predominantly to life science students at undergraduate and master's levels. Our cohorts include some students with an excellent grasp of mathematics (usually in degree schemes such as genetics or population health), but also a significant minority who are much less comfortable with mathematics.

While some statistical concepts are addressed in other modules, the only core statistics teaching is a 10-week module, with a syllabus covering basic concepts, hypothesis testing and practical data analysis. Contact time has varied over the years, but it is usually about 10 h of lectures and 10–20 h of practical exercises.

2.2 Our Teaching Journey

When I first took over as coordinator for this sole (core) undergraduate statistics module (having previously had a non-teaching research role), I found it to be very traditional, in the sense that it focussed predominantly on theory and calculation. Most exercises consisted of manually performing statistical tests (such as ANOVA and regression) in Excel; student feedback was very poor. When I asked former students about the module, they admitted that "nobody knew what they were doing" and they "just copied someone else".

This approach to teaching statistics conflicted with my own instincts and I adopted a new, much more practical, style of teaching. However, all my teaching is carried out in collaboration with a senior colleague and so it took several years of slow evolution before I felt confident enough to fully revise the syllabus in the way I envisioned.

The key feature of our style of teaching is to begin with an attempt to teach core concepts and how a hypothesis test works without the use of mathematics, except for simple addition. We do then move on to the predictable staples of common statistical tests but try to avoid any dependence on equations. It therefore came as a relief when I read the GAISE recommendations [8] as they were aligned very neatly with the direction I had been directing our teaching for some time.

The module currently starts by introducing the concept of, and concepts behind, hypothesis testing. We then show how this knowledge relates to the tools available in analysis packages (SPSS in our case). We cover chi-square, ANOVA and regression individually, and then we extend this knowledge to general linear models and finish with logistic regression and model building. The intention is to provide students with sufficient knowledge to allow them to analyse data by themselves in their final year dissertation.

3 Teaching Without Maths

3.1 Statistics and Maths

We are clearly not the first statistics teachers to reduce the focus on theory. Indeed, books have been written on precisely this subject [10], although more traditional methods, which focus on the "cogs in the mechanism" rather than the principles, do seem to persist with teachers of statistics being "slow to change" [11]. Working in relative isolation in a school of medicine, I do not often get the chance to exchange ideas with colleagues in a similar position. In those discussions I have had with staff in other departments, I have found they generally favour a more mathematical flavour, teaching the students recipes to follow rather than something more transferrable. So, while my experiment is hardly unique, I still feel it is worth describing our approach and our reasons for using it.

3.2   Why Avoid Equations?

An obvious reason to avoid mathematical content is that it is simply not popular with a significant minority of the cohort, who are easily identified by their pained expressions at the start of the first lecture. To see the (obvious) relief on those same faces, an hour later, makes it easy to justify our approach.

We do believe that the mathematical content is worth having, and that there is a limit to the depth of understanding you can achieve without a technical understanding of the calculations. However, we also recognise that we are charged with teaching the only statistics course that many of these students will ever take and we have only 10 weeks to cover a vast and conceptually challenging subject to a mixed ability group. Teaching maths would not only alienate many of our students, but it would also rob us of the time we need to teach other aspects we feel are even more crucial.

Many of our students have studied statistics previously at A level or even at undergraduate level. Almost without exception, they admit that, despite passing the course, they have very little understanding of it. This is hardly surprising given that do not revisit or apply the material after finishing the course. We cannot hope to complete their statistical education in one, short, service course and so aim (primarily) to give them a solid conceptual framework on which to build, should they need to in their future careers.

4   Statistics Using Random Sampling

4.1   Starting with the Core Concepts

The primary challenge we face is helping students understand how hypothesis testing really works. Despite many having completed previous statistics courses, we rarely find a student who can correctly explain what a p-value is. Despite the fact that there is an ongoing and vigorous debate on their value, hypothesis testing and p-values are important concepts and if our students go into academic research they are not going to be able to avoid these two terms.

We tackle this problem head-on by starting the course with a workshop in which they construct a hypothesis test and calculate a p-value using random sampling, a task they (almost all) manage to complete with only gentle support from the teaching staff.

4.2   Structure of the Session

The first session of the course starts with the teaching staff organising students into small groups and handing each group a pack of 10 dice, each with 3 blanks (i.e., "patient deteriorates") and 3 coloured sides (i.e., "patient improves"). We describe a scenario in which a clinical trial of an experimental drug saw eight out of the ten participants registering an improvement in some metric of their symptoms, making it clear we were keeping things simple by only considering these two outcomes. We chose to use dice, rather than coins, as the process of rolling dice was quicker, less noisy and more satisfying. However a coin (or coins) could easily be used instead. We then explain to them that we want them to use the tools we have given them (the dice) to construct a test to assess the claim by the makers of this drug that the drug is effective.

In most groups a careful inspection of the dice follows and an earnest discussion of their purpose. Dice are rolled and more discussions follow. We move among the groups, asking them what they think the purpose of the dice is and what the numbers they are generating represent. The first important step is for them to recognise that individual dice represent individual patients whose outcome is determined by chance. The second key step is to see that rolling all 10 dice at once is necessary to shed light on the likelihood of the observed outcome. In some cases they make both of these steps by themselves but often they will need a little prompting.

For instance, if a group seems stuck then we may remind them that patients can only improve or deteriorate (no ties allowed) and that this may, or may not, be affected by the drug treatment. Or we might question them on what they think the dice might represent. With the odd prompt such as these, almost all groups will begin using all 10 dice to simulate the drug trial under the null hypothesis (although we do not use this term yet), where patient outcomes are not affected by the drug treatment and improvement and deterioration are equally likely.

After 10–20 simulations most groups are happy to conclude that an 80% success rate is not impossible, but is very unlikely to occur as a result of chance. If pushed they conclude that the drug is probably effective.

4.3   Building on the Basics

Once all groups have reached this, or indeed any, conclusion we spend the second half of the session going over the process they have conducted, explaining the key concepts along the way.

We start by explaining what the null hypothesis and test statistic means in this context. We then help them see that they have generated a (crude) distribution for this test statistic under the null hypothesis and finish by explaining where the p-value fits in.

The final step is to go over the process again, showing them that they are performing a binomial test, and working through output generated by SPSS so they can see that they have performed a real statistical test, and understood it.

5   The Response from the Students

The students mostly seem to enjoy the session and it serves several useful purposes. It helps to introduce the basic concepts without getting bogged down in technical terms or equations. It also acts as an ice-breaker, since we are able to easily draw them into conversation, making it easier for them to approach us with their questions later on.

The feedback we get is very positive, from both ends of the ability spectrum. This session was singled out for praise by some students for easing them in with "an interesting but relevant activity" and for making "difficult concepts simple and easy to understand". Most satisfying is the response we get from the most anxious students and those with no prior experience of statistics who are visibly relieved by the tone we have struck from the outset. In our end of module feedback, we reliably get a significant number of students leaving comments to the effect that they had expected to hate the module but ended up surprising themselves with how much they enjoyed it.

More surprising, and perhaps even more satisfying, is the response we get from students who have studied statistics before. At the end of every session, almost without fail, one or two students will approach us and tell us they have learned more in this session than in their entire (previous) course and that now it all makes sense. The only significant complaint about the module comes from a subgroup of students who would have liked more theory.

Possibly the most convincing defence I can put forward for the way we have chosen to teach is the change in students' attitudes towards the subject. Tired of teaching only basic statistics, I consulted some of our students on whether they would be interested in an advanced statistics module in their final year. The response from the students, and from the programme directors, was positive so I developed a new, final year, module which covers a range of more advanced methods (e.g., generalised linear models, survival analysis, mixed models) with a similarly practical emphasis on concepts over theory. In the first year it ran, nearly 50% of the entire cohort (of 250 students) opted to take the module, leaving me speechless (and very busy).

6   Conclusions

My early experience of basic statistics was rather different to that of most undergraduates in that I was largely self-taught during a previous career as an analyst in a social research company. On arriving in university, I was (re)taught statistics simultaneously in a mathematics department and a business school. One department chose a dry, theoretical approach dominated by mathematics, while the other took a practical, interactive and applied approach. The difference between the two was stark and left me in no doubt how the subject should be taught.

My teaching team believe in engaging students, with an emphasis on those who might otherwise be turned off the subject. It has been a tricky balance, trying to satisfy both ends of the ability spectrum, a gap that I do not think we will ever be able to fully bridge.

It speaks volumes that we can receive very positive student feedback (relative to other modules) despite delivering a core module in a subject students have a reputation for disliking. We feel our approach that it is in line with current recommendations and, if nothing else, convinces the students that statistics is not something impenetrable and terrifying.

References

1. Onwuegbuzie AJ, Wilson VA. Statistics anxiety: nature, etiology, antecedents, effects, and treatments--a comprehensive review of the literature. Teach High Educ. 2003;8(2):195–209.
2. Dunn DS, Smith RA, Beins BC. Overview: best practices for teaching statistics and research methods in the behavioral sciences. London: Routledge; 2007.
3. Bessant KC. Instructional design and the development of statistical literacy. Teach Sociol. 1992;20(2):143–9.
4. Bromage A, Pierce S, Reader T, Compton L. Teaching statistics to non-specialists: challenges and strategies for success. J Furth High Educ. 2021;46:61. https://doi.org/10.1080/0309877X.2021.1879744.
5. Levpušček MP, Cukon M. That old devil called 'statistics': statistics anxiety in university students and related factors. Cent Educ Pol Stud J. 2021;12:1.
6. Hulsizer MR, Woolf LM. A guide to teaching statistics: innovations and best practices. New York, NY: John Wiley & Sons; 2009.
7. Franklin C, Kader G, Mewborn D, Moreno J, Peck R, Perry M, Scheaffer R. Guidelines for assessment and instruction in statistics education (GAISE) report; n.d.
8. Carver R, Everson M, Gabrosek J, Horton N, Lock R, Mocko M, Rossman A, Roswell GH, Velleman P, Witmer J, Wood B. Guidelines for assessment and instruction in statistics education (GAISE) college report; 2016.
9. Hayat MJ. Guidelines for assessment and instruction in Statistics Education (GAISE): extending GAISE into nursing education. J Nurs Educ. 2014;53(4):192–8.
10. Dancey CP, Reidy J. Statistics without Maths for psychology. London: Pearson Education; 2007.
11. Cobb GW. The introductory statistics course: a Ptolemaic curriculum? Technol Innov Stat Educ. 2007;1(1):1.

COVID-19: Online Not Distant—MSc Students' Feedback on an Alternative Approach to Teaching 'Research Methods and Introduction to Statistics' at UCL Queen Square Institute of Neurology

Saiful Islam, Saiam Ahmed, Rosamund Greiner,
Shah-Jalal Sarker, Mifuyu Akasaki, Masuda Khanom,
David Blundred, Alessandro Cozzi-Lepri,
and Yasna Palmeiro-Silva

1 Introduction

The COVID-19 pandemic necessitated a swift change in the delivery of university education, with teaching and learning moved online, and a range of technologies and approaches employed to engage students in remote learning. Long before the pandemic, those delivering health and medical education had added their voices to the broader call for change in the way university education was delivered. They called for a reorientation away from teacher-centred methods and towards learner-centred approaches for engagement [1–5]. This movement stresses that learning is enhanced by reflection, group discussion, and application to real-life scenarios [6].

Learner-centred approaches include the flipped classroom approach (FC) and blended learning. FC, broadly defined, is an approach to education that reverses the traditional method of education delivery. In a traditional approach, information is transmitted to students in a group setting, typically by lectures, with

S. Islam · S.-J. Sarker · M. Khanom · D. Blundred
Institute of Neurology, University College London, London, UK

S. Ahmed
MRC Clinical Trials Unit, University College London, London, UK

R. Greiner · A. Cozzi-Lepri · Y. Palmeiro-Silva (✉)
Institute for Global Health, University College London, London, UK
e-mail: yasna.palmeiro.18@ucl.ac.uk

M. Akasaki
Institute of Education, University College London, London, UK

recapping, consolidation, and application to be undertaken individually through homework, quizzes, and problem sheets [7]. In FC, students are provided with content to read, watch, or listen individually in their own schedule, leaving class time free for recapping, consolidation, and application through interaction, discussion, and debate amongst students and educators [8]. FC in the university setting has been found to improve student performance, motivation, engagement, and satisfaction [9–13]. FC has been applied in medical education and resulted in improved student performance and satisfaction [11, 14, 15]. Specifically, the use of in-class quizzes to consolidate learning increased the effectiveness of FC in medical education [11].

In its original use, the term 'blended learning' referred to the combining of in-person instruction with the use of e-learning methods [16]. However, the definition of blended learning has evolved over time as technology has developed, and it has come to refer to a broad range of educational approaches that have in common the use of a number of complementary teaching and learning methods [16]. These can consist of a mixture of online and in-person synchronous activities, in which all participants and instructors are present in the same (online) space at the same time, and online asynchronous activities, which instructors set for learners to complete at their own pace [17]. Traditional blended learning, including in-person instruction, has been applied in medical education and resulted in better student performance than traditional approaches [17]. For much of the COVID-19 pandemic, in-person learning was rendered impossible. Due to this, blended learning approaches have been modified to incorporate a range of online teaching and learning methods including pre-recorded lectures, discussions, forums, and quizzes [18, 19]. We adopt this definition of modified blended learning that has been developed specifically in the context of COVID-19 and limitations on conducting in-person instruction [18, 19]. FC can be used in conjunction with a modified blended learning approach to offer a flexible, learner-centred, and effective approach to education [18]. Unfortunately, few studies have investigated student satisfaction with fully online flipped and modified blended learning in medical education.

Due to COVID-19 restrictions, the delivery of the *'Research Methods and Introduction to Statistics'* (RMIS) module at the UCL Queen Square Institute of Neurology students was converted from face-to-face teaching to online delivery. The objective of this article was to describe the students' feedback on the teaching and learning methods used in this modified blended approach.

2   Materials and Methods

2.1   Setting

This study took place at the UCL Queen Square Institute of Neurology, specifically analysing the Master's module *'Research Methods and Introduction to Statistics'*. This module was delivered in the first term of the 2020–2021 academic period and was 12 weeks long.

2.2 Participants

A total of 118 students were included in this study. Students were from a wide range of postgraduate programmes from UCL Queen Square Institute of Neurology including Clinical Neuroscience MSc, Stroke Medicine MRes, Brain and Mind Sciences MSc, Dementia (Neuroscience) MSc and Advanced Neuroimaging MRes.

2.3 Module Design and Delivery

We developed a modified blended approach that incorporated FC and involved both synchronous and asynchronous online learning activities. Every week, students were engaged in both independent and interactive learning. For each week, there were four components: preparation, recap, a synchronous session, and reference materials. Table 1 shows a summary of all components considered in this new approach, demonstrating the flipped classroom approach.

The module has been designed in line with a modified blended learning approach incorporating the flipped classroom and online delivery. Upon enrolling in the course, all students were provided with a suggested timeline for completion of the learning activities in each week. The provided timeline made clear to students when lecture slides and videos would be available and when they should attend synchronous sessions. We expected that after completion of the preparation and recap materials, students may still not have a full understanding of the topic, especially as many of them do not have previous research experience or education in research methods or statistical analysis. The synchronous sessions and Moodle forums provided essential support to help students develop a good understanding of each topic and to prepare them for their module assessment. The module assessment was planned to follow a similar format to that of the recap quizzes.

Asynchronous, Independent Activities: Preparation and Recap

Preparation and recap materials were released weekly, and it was suggested to students that they should complete their reading, watching lecture videos and attempt the quiz on Moodle at least once before the Q&A session for that topic. In weeks when the quiz was replaced with a Stata workshop, students were provided with datasets and problem sheets to attempt before the workshop.

Synchronous, Interactive Sessions: Q&As and Workshops

There were a total of ten synchronous sessions: eight weekly 1-h Q&A sessions and two 3-h Stata workshops. Each week, students were randomly allocated into eight groups, resulting in a class ratio of approximately 1 tutor to 15 students. Being in small groups and having access to a tutor for live interaction gave students the opportunity to ask any questions they had about the topic and interact with their peers through discussion of the topic.

To ensure that a consistent learning experience was offered by all tutors and across all weeks, the tutors followed a standard structure for the Q&A sessions.

Table 1 Components of modified blended learning for RMIS 2020

Component	Contents	Frequency	Delivery mode	Style of engagement	Delivery description
Preparation materials	Notes on key concepts and terminology	Weekly	Asynchronous	Independent	Uploaded to Moodle in advance
	Lecture slides	Weekly	Asynchronous	Independent	Uploaded to Moodle in advance
	Lecture videos	Weekly	Asynchronous	Independent	Uploaded to Moodle in advance
Recap materials	Multiple choice quiz	Weekly (weeks 1–4 and 6–9)	Asynchronous	Independent	Uploaded to Moodle—to be attempted after completing preparation materials
	Stata datasets for students to investigate, exercises to attempt	Twice termly (weeks 5 and 10)	Asynchronous	Independent	Uploaded to Moodle—to be attempted after completing preparation materials
Synchronous sessions	Questions and answers	Weekly (weeks 1–4 and 6–9)	Synchronous	Interactive	Students are split into small groups and join Zoom session with a module tutor
	Stata workshop	Twice termly (weeks 5 and 10)	Synchronous	Interactive	Students are split into small groups and join a Zoom session with a module tutor
Reference materials	Synchronous session recording	Weekly	Asynchronous	Independent	Uploaded to Moodle after the session
	Moodle forum	Available throughout module delivery	Asynchronous	Interactive	On Moodle, answered by tutors throughout the module

Note: Synchronous activities covered live and simultaneous sessions; asynchronous activities were self-paced learning activities; independent refers to activities that were completed by each student, and interactive activities occurred when students worked together

Each Q&A session began with 10 min for the tutor to introduce themselves, '*break the ice*' and allow any students running late to join the class. Tutors spent 15 min to recap key concepts from the lecture topic of the week. The following 25 min were used to review the quiz questions that students had attempted in advance. The tutors facilitated discussions and asked students which they thought was the correct answer, then revealed the correct answer and provided a brief explanation given the responses. The remaining 10 min were used to clarify any questions students might have about the topic, including highlighting questions that students had raised on the Moodle forum.

In weeks 5 and 10, rather than attempting a quiz, students were asked to download sample datasets and attempt exercises using Stata prior to attending the workshop. The synchronous workshop consisted of tutors providing a live demonstration of selected exercises in Stata with input from the students based on their previous attempts.

Asynchronous, Interactive Activities: Discussion Forums

A Moodle forum called '*Your questions answered*' was created. Students were encouraged to post questions related to the module content to the forum, which was visible to all enrolled students and tutors. Tutors monitored the forum and responded to all questions promptly, providing an explanation of the topic along with links to additional resources if appropriate.

All synchronous sessions were recorded so that if any students were unable to attend, they could catch up at their own schedule. The Moodle forum enabled students to ask questions on previous week's topics if they had missed a synchronous session, and these were answered on the forum by tutors and if necessary, addressed in the next Q&A.

Moodle Page

From students' feedback in previous years, we found that students had struggled to find the materials on Moodle. This, combined with the move to online delivery of the module, necessitated a re-organisation of the Moodle page to be more accessible for online learning. A drop-down tab was created for each week containing all preparation materials in order, the recap materials, and a link to the week's synchronous session. After each synchronous session, the recordings were uploaded under the week's tab. A survey named '*your feedback*' was added to each week after the synchronous session had concluded. Table 2 shows the organisation of resources on the Moodle page for the module from enrolment to the final exam.

2.4   Collection of Feedback

Students were encouraged to give feedback after each Q&A session and assured that their participation and responses were anonymous. Four questions were Likert-type with five alternatives (1 = 'very bad' to 5 = 'excellent') and two questions were open. The questions were the following:

1. Was the session(s) fit for purpose (what you needed)?
2. Was the Q&A mapped to the original lecture slides and videos?
3. Was the Q&A session delivered at the right speed?

Table 2 Organisation of resources on Moodle

Timeline	Moodle activity
Between enrolment and Week 1	Ice breaker activities
	Pre-reading material organised by student level (beginner or advanced)
Whole module	'Your questions answered' forum
	Synchronous session group allocations and joining instructions
Week 1–4: Lectures	Preparation materials (lecture) in order
	Recap materials—Moodle quiz
	Link for synchronous session (Q&A), with date and time
	Reference materials
	'Your feedback' survey
Week 5: Stata Workshop 1	Preparation materials (workshop 1) in order
	Recap materials—Stata exercises
	Link for synchronous session (workshop 1), with date and time
	Reference materials
	'Your feedback' survey
Week 6–9: Lectures	As in week 1–4: Lectures
Week 10: Stata Workshop 2	As in Stata Workshop 1
Week 11: Revision	Revision materials
Week 12: Exam	Exam materials

4. Were the discussions and interactive parts of the session useful?
5. Please suggest one point of improvement for the next lecture (open)
6. Any other comments about any other aspect of this session (open)

2.5 Statistical Analysis

Descriptive analyses of the feedback during the flipped teaching year were conducted and the breakdown of feedback responses by week was considered. Analyses were conducted using the Stata v16 software (StataCorp. 2019 Stata Statistical Software: Release 16 College Station, TX StataCorp LLC).

3 Results

A total of 66 responses were received in the week-by-week longitudinal feedback. The breakdown of the students providing feedback data by week is shown in Table 3. Approximately 15% of those who participated in the survey were active during week 1 and week 2. After week 2 the proportion of students providing feedback dropped to less than 6% of the total feedback providers.

Regarding Likert-type questions, most of them presented positive feedback. Figure 1a–d shows the proportion of students reporting each category. For the question 'Was the session(s) fit for purpose (what you needed)?' the categories 'very good' and 'excellent' increased over time going from 40% and 0% in week S1 to 40% and 60% in week S8, respectively (Fig. 1a). However, in week S7

Table 3 Number and proportion of responses per week

Session	Number of responses (%)
1	20 (16.7%)
2	15 (12.5%)
3	7 (5.8%)
4	4 (3.3%)
Workshop 1	7 (5.8%)
5	2 (1.7%)
6	3 (2.5%)
7	3 (2.5%)
8	5 (4.2%)
Total	66

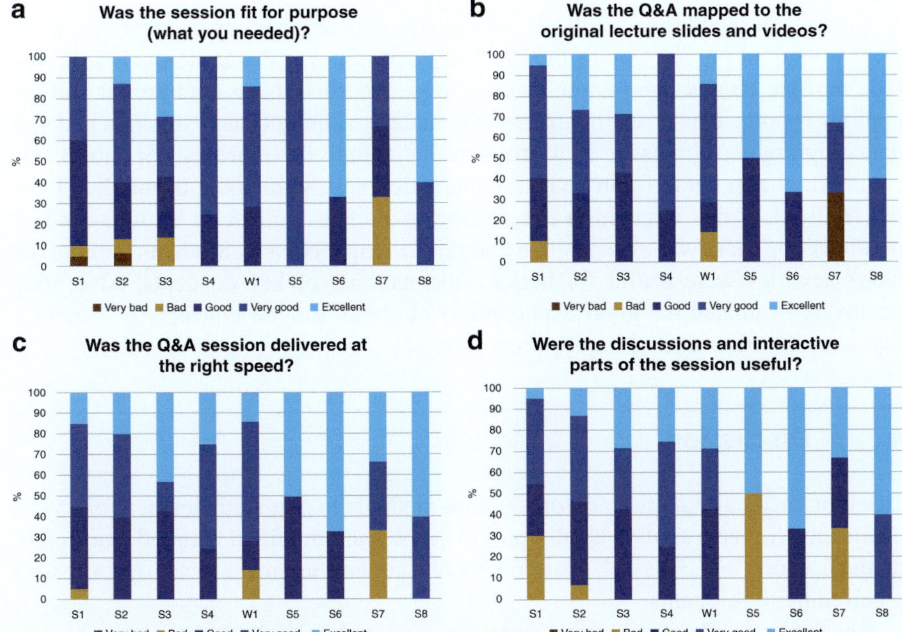

Fig. 1 Distribution of feedback for Likert-type questions. S1: Session 1; S2: Session 2; S3: Session 3; S4: Session 4; S5: Session 5; S6: Session 6; S7: Session 7; S8: Session 8; W1: Workshop 1

(observational study designs), 1/3 of responses were 'bad' but in absolute numbers, this corresponds to only one person.

For the second question, the proportion of students reporting the excellent rank was 5% in S1 vs. 60% in S8 (Fig. 1b). Similar to question 1 in S7, only one person gave 'very bad' feedback.

The third question was about the tutor's ability with delivering sessions and appropriateness of the length and teaching pace. Specifically, we asked: 'Was the Q&A session delivered at the right speed?' (Fig. 1c). For this particular outcome, we found that the proportion of students providing excellent feedback scores also increased from 15% in S1 to 60% in S8. In S7, same pattern is found as in questions 1 and 3.

The last Likert-type question aimed to evaluate the discussions and interactive features of the session (Fig. 1d). The proportion of students reporting excellent scores increased over time (5% in S1 vs. 60% in S8). However, in S5 (correlation and sample size calculations), 50% of responses were 'bad', which correspond to one person.

Overall, the students were satisfied with the appropriateness of sessions regarding the purpose and content, as well as the coherence between the Q&A session and lectures. Conversely, students were less satisfied about the interaction during the Q&A sessions.

In terms of qualitative feedback, which was obtained from open questions, there were important points. One of the most relevant issues is the audio quality of specific past lectures because some of them were recorded using Lecturecast and the audio was poor. This problem affected the understanding of these sessions and students highlighted this issue as a problem that should be improved for future sessions. Another point to improve is the interaction with students. It is not clear from the feedback if this interaction is referring to student–student or lecturer–student relationship. There was also positive and reinforcing feedback. Students thought the Q&A sessions were useful for better understanding of key concepts. Also, they positively evaluated the good explanations of some difficult concepts or concepts that were not clear during the lecture.

4   Discussion

Teaching medical statistics and/or research methods to non-statisticians is always a challenge. We have applied a fully online flipped and modified blended learning in medical education. This self-learning mixed method approach was, overall, well received by the students.

Regarding feedback provided over time, over the 2020–2021 academic year, results were encouraging although more difficult to interpret. Of note, because we posted lecture videos and slides in advance and asked students to listen and read them before the question and answer (Q&A) session, we were able to ask relevant questions such as: 'Was the Q&A session mapped to the original lecture slides and videos?' During the Q&A sessions we also had some interactive parts with the students actively participating asking for clarifications, questions posted on the Moodle forum which remained unsolved or providing a live demo of Stata coding for verification. Overall, the proportion of students reporting higher level of satisfaction to the Q&A sessions showed an increase over time. However, there are several limitations to this analysis. Firstly, the analysis assumes independence of the feedback responses while it is possible that some could have been sent by the same student. As the feedback was provided anonymously, we could not control for extra-correlation between feedback received from the same student. Secondly, there was likely selection bias as we estimate that approximately 56% of the total class participated in the survey so the analysed sample is likely not to be representative of the whole class. Thirdly, there was large attrition over time with only 6% of the feedback providers who kept sending feedback past the second week of the course. This high rate of drop-out could have been caused by general study fatigue rather than loss of interest for the subject, although we cannot rule out that only students who were particularly enjoying the course were those providing feedback until the end.

The weekly feedback, especially from the open questions, was used by tutors to act upon and improve the style of teaching. Although the practice of asking students' feedback is not exclusive to these online approaches, it allows and supports continue and timely improvement, as well as the maintenance of quality and standards in higher education [20]. One explanation why positive feedback was increasing over time might be related to this improvement; therefore, weekly feedback is recommended as a usual practice.

As Q&A sessions were recorded by all tutors, providing the recordings enabled students to look at them from different groups—each tutor had a different way of explaining the material and quiz answers, so students could seek an alternate explanation to widen their knowledge understanding. This flexibility helped students to understand the topic more effectively.

As this mode of delivering had to be set in a quite short time, there were several areas of teaching and learning that should be improved. Better recordings and clear materials are needed to improve the understanding of the content. Also, protected learning environments and spaces where students could ask and clarify key concepts are crucial for learning and improving students' confidence, especially in the field of medical statistics. One area to consider in future versions of this module is the measurement and monitoring of the degree of use of the materials by students. This could be a key indicator of students' engagement and the module, as well as a potential predictor of students' performance. Also, it could highlight areas or specific materials that need improvements or further development.

In conclusion, the modified blended learning seems to be a complementary alternative to the traditional methods for teaching medical statistics and/or research methods to non-statisticians. Various techniques including Q&A, workshops, and an online forum (like Moodle) are important to enhance student learning. Specifically, interactions help students understand the topic they found challenging in pre-reading.

References

1. McLean M, Gibbs TJ. Learner-centred medical education: improved learning or increased stress? Educ Health. 2009;22:287.
2. Mehta NB, Hull AL, Young JB, Stoller JK. Just imagine: new paradigms for medical education. Acad Med. 2013;88:1418–23. https://doi.org/10.1097/ACM.0b013e3182a36a07.
3. Prober CG, Heath C. Lecture halls without lectures — a proposal for medical education. N Engl J Med. 2012;366:1657–9. https://doi.org/10.1056/NEJMp1202451.

4. Prober CG, Khan S. Medical education reimagined: a call to action. Acad Med. 2013;88:1407–10. https://doi.org/10.1097/ACM.0b013e3182a368bd.

5. Spencer JA, Jordan RK. Learner centred approaches in medical education. BMJ. 1999;318:1280–3. https://doi.org/10.1136/bmj.318.7193.1280.

6. van der Vleuten CPM, Driessen EW. What would happen to education if we take education evidence seriously? Perspect Med Educ. 2014;3:222–32. https://doi.org/10.1007/s40037-014-0129-9.

7. Walker Z, Tan D, Klimplová L, Bicen H. An introduction to flipping the classroom. In: Walker Z, Tan D, Koh NK, editors. Flipped classrooms with diverse learners: international perspectives. Singapore: Springer; 2020. p. 3–15. https://doi.org/10.1007/978-981-15-4171-1_1. Springer Texts in Education.

8. O'Shea PM. Flipped learning at the university level. In: Walker Z, Tan D, Koh NK, editors. Flipped classrooms with diverse learners: international perspectives. Singapore: Springer; 2020. p. 171–81. https://doi.org/10.1007/978-981-15-4171-1_10. Springer Texts in Education.

9. Colomo-Magaña E, Soto-Varela R, Ruiz-Palmero J, Gómez-García M. University students' perception of the usefulness of the flipped classroom methodology. Educ Sci. 2020;10:275. https://doi.org/10.3390/educsci10100275.

10. Doo MY, Bonk CJ. The effects of self-efficacy, self-regulation and social presence on learning engagement in a large university class using flipped Learning. J Comput Assist Learn. 2020;36:997–1010. https://doi.org/10.1111/jcal.12455.

11. Hew KF, Lo CK. Flipped classroom improves student learning in health professions education: a meta-analysis. BMC Med Educ. 2018;18:38. https://doi.org/10.1186/s12909-018-1144-z.

12. Jdaitawi M. Does flipped learning promote positive emotions in science education? A comparison between traditional and flipped classroom approaches. Electron J E-Learn. 2020;18:516–24. https://doi.org/10.34190/JEL.18.6.004.

13. Meyliana, Sablan B, Surjandy, Hidayanto AN. Flipped learning effect on classroom engagement and outcomes in university information systems class. Educ Inf Technol. 2021;27:3341. https://doi.org/10.1007/s10639-021-10723-9.

14. Anderson HG, Frazier L, Anderson SL, Stanton R, Gillette C, Broedel-Zaugg K, Yingling K. Comparison of pharmaceutical calculations learning outcomes achieved within a traditional lecture or flipped classroom andragogy. Am J Pharm Educ. 2017;81:70A, 70B, 70C, 70D, 70E, 70F, 70G, 70H, 70I.

15. Hu X, Zhang H, Song Y, Wu C, Yang Q, Shi Z, Zhang X, Chen W. Implementation of flipped classroom combined with problem-based learning: an approach to promote learning about hyperthyroidism in the endocrinology internship. BMC Med Educ. 2019;19:290. https://doi.org/10.1186/s12909-019-1714-8.

16. Singh H. Building effective blended learning programs. In: Challenges and opportunities for the global implementation of e-learning frameworks. New Delhi: IGI Global; 2021. p. 15–23. https://doi.org/10.4018/978-1-7998-7607-6.ch002.

17. Vallée A, Blacher J, Cariou A, Sorbets E. Blended learning compared to traditional learning in medical education: systematic review and meta-analysis. J Med Internet Res. 2020;22:e16504. https://doi.org/10.2196/16504.

18. Nerantzi C. The use of peer instruction and flipped learning to support flexible blended learning during and after the COVID-19 pandemic. Int J Manag Appl Res. 2020;7:184–95.

19. Ożadowicz A. Modified blended learning in engineering higher education during the COVID-19 lockdown—building automation courses case study. Educ Sci. 2020;10:292. https://doi.org/10.3390/educsci10100292.

20. Brennan J, Williams R. Collecting and using student feedback - a guide to good practice. York: Learning and Teaching Support Network; 2004.

Common Misconceptions of Online Statistics Teaching

Eirini Koutoumanou

1   Introduction

Statistics is a topic feared by many, both at student and professional working level [1]. Those who are not studying statistics as their main subject area are usually taught fundamental statistical concepts in one or two modules over the course of one or two academic terms. Professionals at a working level may turn either to self-paced material, in the form of videos, books and online interactive learning platforms, or to face-to-face (f2f) or online webinars/short courses, or a combination of all the above [2, 3].

The Centre for Applied Statistics Courses (CASC) is a team of statisticians based at the Population, Policy and Practice Research and Teaching Department (PPP) of the Great Ormond Street Institute of Child Health (GOSICH) department of University College London (UCL). The aim of the centre is to offer statistics training to students, researchers and professionals from all backgrounds and institutions in the form of short courses [4].

Effective short courses on statistics, as well as other topics, should have clear and realistic educational objectives that can be met within the (short) duration of the course. These should be transparent and clearly presented to all participants prior to attendance [5]. Due to their short duration, they should lay solid and relevant foundations right from the start to enable participants to settle in swiftly and become familiar with the terminology used and the style of teaching. Numerical and formulaic expressions should ideally be avoided, or at least be kept to a minimum or included in optional appendices/supplementary materials, within areas such as statistics. These topics can be hard for large portions of audiences who do not have a

E. Koutoumanou (✉)
Great Ormond Street Institute of Child Health, Population, Policy and Practice Research and Teaching Department, University College London, London, UK
e-mail: e.koutoumanou@ucl.ac.uk

mathematics or statistics background (such as those that often attend the CASC courses), and they might only benefit a small minority of such participants. The language used should be simple and understandable to all with strong emphasis on the applicability and interpretability of statistical results and output [6].

Delivery of online classes emerged as a necessity for all levels of the education sector in March 2020 — some educators had almost no time to prepare. The debate about the benefits and challenges of online learning, however, has been active for a while now [7]. Over the last couple of years though, this debate has seen a major expansion with the vast majority of teaching staff experiencing online teaching closely, and possibly for the first time.

In this chapter, I reflect retrospectively on CASC's experience during the switch to online short-course teaching and present evidence that have led us disprove myths about online teaching that many of us at CASC believed previously. The latter is presented in the form of our top 10 misconceptions of online statistics teaching.

2   The Operating Model of CASC

Since its inception in 2008, CASC has been formed of 2–5 members of teaching staff and supported by an administrator. Each member of CASC's teaching team is responsible for leading their own short course(s) and share the teaching of other longer (short or credit bearing) courses.

CASC is responsible for the credit bearing statistics module offered to nearly all taught postgraduate and intercalated BSc students at GOSICH (average of 85 students per year) and contributes to statistics modules of other UCL departments too (average of 105 students annually). CASC also delivers short courses on 19 different statistical topics and has done so regularly since 2008. Topics include the introduction to statistics, various regression analysis techniques, sample-size estimation, missing data techniques, Bayesian analysis and more [4]. Hands-on training for statistical software/languages is also offered, specifically in R, SPSS, Stata and Matlab, as summarised in Table 1.

Class sizes range between 15 and 35 participants depending on the popularity of the content taught but are always capped to 40 participants with at least two teaching staff available to ensure the best learning experience for our audience. Such ratios of teaching staff and participants have proven suitable, based on our experience, to ensure that both participants and lecturers are well supported. Computer-based courses are restricted specifically by the classroom sizes available which often cannot host more than 30/40 computers.

The duration of the short courses ranges from 6.5 to 30 contact hours over the course of 1 or 2 half days (9.30 am to 1 pm) to a full teaching week (9.30 am to 5 pm daily). More specifically, CASC runs an average of 60 courses per year (including 2–4 repeats of individual courses) equaling an approximate total of 400 contact hours and an average target audience of approximately 1200 participants.

Table 1 Titles and syllabus of courses offered by CASC

	Course title	Syllabus
1	Introduction to research methods and statistics	Study design, descriptive statistics, statistical inference/significance (CI, p-values, parametric/non-parametric tests), bootstrapping, linear regression
2	Introduction to regression analysis	Types of regression, correlation, simple and multiple models, interpretation, inference, prediction, diagnostics, model validity
3	Introduction to logistic regression	Binary variables, transformations, odds and odds ratios, simple and multiple logistic models, interpretation, inference, prediction, diagnostics, model validity
4	Introduction to time-to-event data analysis	Time-to-event outcome variables, hazard ratio, survival function, log-rank test, simple and multiple time-to-event models, interpretation, inference, prediction, diagnostics, model validity
5	Introduction to Poisson regression	Count data, rates, rate ratios, simple and multiple Poisson models, interpretation, inference, prediction, diagnostics, model validity
6	Introduction to dealing with missing data	Types of missing data, listwise and pairwise deletion, mean and regression imputation, simple and multiple imputation, sensitivity analysis
7	Overview of regressions with R	Types of regression, application of linear, logistic, ordinal logistic, time-to-event, Poisson, negative binomial regressions in R, interpretation, inference, prediction, diagnostics, model validity
8	Introduction to R	R interface, objects, functions, data frames, descriptive statistics, graphs, statistical tests, packages
9	Introduction to SPSS	SPSS interface, data entry, data editing, descriptive statistics, graphs, statistical tests, linear regression
10	Introduction to Stata	Stata interface, commands, data entry, variables, descriptive statistics, graphs, statistical tests
11	Further topics in R	Dataset merging, loops, conditional statements, apply and write functions
12	Introduction to data analysis using MATLAB	MATLAB interface, objects, functions, data entry, descriptive statistics, graphs, statistical tests, linear regression
13	Sample-size estimation and power calculations	Precision, power, calculations for: single mean, difference in means, single proportion, difference in proportions, single rate, difference in rates, hazard ratio
14	Chi-square and beyond for 2×2 tables	Proportions, percentages, odds, odds ratios, relative risk, chi-square test, Fisher's test, sensitivity, specificity, agreement
15	Analysis of variance (ANOVA)/general linear models (GLM) in SPSS	Mean sum of squares, F test, one way and factorial ANOVA, repeated measures and multivariate ANOVA
16	Critical appraisal	Research design, types of variables, descriptive statistics, statistical inference, types of significance tests and regressions, appraisal guidelines
17	Assessing measurement reliability and validity	Internal, external and measurement validity, diagnostic testing, limits of agreement, internal consistency
18	Introduction to Bayesian analysis	Frequentist analysis, prior, posterior and likelihood functions, Bayes theorem, credible and highest posterior density intervals, Bayes factor
19	Introduction to meta-analysis	Systematic reviews, Cochrane, pooled effects, fixed and random effects, heterogeneity, publication bias

2.1 Pre-COVID19 Era

Before the COVID-19 pandemic started to affect everyday life (and certainly prior to March 2020), courses were conducted conventionally in a classroom or computer room at UCL. Teaching days in the pre-COVID era would always commence at 10 am and end at 4.30 pm with an hour's lunch break halfway through each session and regular coffee breaks.

Teaching involved slide presentations and plenty of hands-on practicals (either pen and paper or computer-based exercises). Additionally, live voting was incorporated in the slide presentation with results appearing instantaneously on the slides along with live annotations made using a digital pen.

An informal, friendly environment was nurtured right from the start of each course with the aim of making participants feel comfortable and confident to voice any confusion and ask questions. Evaluation forms were sent to all participants at the end of each course with rating questions ranging from 0 (poor) to 5 (excellent) about various aspects of the course. A common question across all courses/forms related to the overall quality of education offered by CASC. Average feedback ratings for this question have been consistently between 4 and 5 over the years. Many short course participants return for more training of further relevance to them and a large portion of first-time attendees register on the basis of recommendations from previous delegates — both evident by a growing list of CASC newsletter subscribers, growing from 0 to just over 9000 in the past 14 years with no formal marketing campaign other than word-of-mouth.

2.2 Post-COVID19 Era

The CASC team ran its first online teaching session with next to no preparation on 16 March 2020. We were halfway through a f2f'Introduction to Research Methods and Statistics' short course, which overnight was switched to an online course delivered using Microsoft Teams [8]. We felt uncertain at the time about the course experience we could offer our students/short course participants; confused about the choices we had to make regarding technical aspects of our setup, equipment we might require and more; and reluctant about embarking on a journey into unfamiliar territory [9].

The next few weeks were a steep learning curve for all of us at CASC, with several more courses taking place as per our original timetable, which we did not want to disrupt with minimal notice for our participants, and some of our initial feelings started to be replaced by a different set of emotions: we felt empowered being able to reach a far wider audience than ever before; we reflected on participants' feedback (variety of comments listed below) on how online classes have made attendance to our courses possible (since previously some had struggled with commuting, caring responsibilities and costs); we felt motivated to explore new avenues for delivering high-quality online statistics training (although we were still somewhat uncertain about the breadth of options available to us); but, let us not forget, there were the frustrations caused by IT failures too.

Examples of relevant quotes from students are listed below (such quotes will appear in other parts of the chapter too; they have all been submitted by short course participants over the last 2 years on our anonymised feedback forms):

Loved that it could be done online, otherwise I would have to travel plus all the expenses associated.

… the Zoom on line course are a major milestone. I used to do a course once a year, now I can manage to do 2 or 3 courses every year.

One of the first changes introduced to our operating model, based on feedback from our participants as well as on our own experience, was the duration and timings of our classes. Courses normally delivered on a single day were broken up into two halves; two morning sessions from 9.30 am to 1 pm. This broke up the previously intense and draining setup of staring at a screen for 5–6 h (10 am to 4.30 pm).

By the end of July 2021, we had completed and delivered 24 further online short courses and each one of them played an important role in teaching us how to better plan and conduct online teaching. Feedback from our participants was invaluable in giving us tips for improving delivery from the students' perspective. All these lessons were carried forward in the scheduling of courses for 2021–2022. These reflections are outlined in the next section in the form of 'CASC's top 10 Misconceptions about Online Statistics Courses'.

3   Top 10 Common Misconceptions About Online Statistics Courses

The misconceptions have been categorised into three sections: infrastructure and equipment, teaching methods and classroom environment and, finally, planning ahead (Fig. 1). Following this section, we conclude with some overall remarks and our thoughts on where the future of short statistics courses is heading.

Several of these misconceptions, especially in the first category, are somewhat dependent upon decisions made by the institution hosting the courses and/or the financial structure operating with respect to the immediate teaching team (or the department within which the teaching team/staff are based). Our experience at CASC has therefore been very much dependent on UCL's infrastructure, software subscriptions and our immediate team's financial discretion — all of which might vary significantly amongst different institutions/teams.

3.1   Infrastructure/Equipment

Misconception 1 The IT infrastructure will not be good enough for an online class of 30+ students.

Infrastructure / Equipment

- The IT infrastructure will not be good enough for an online class of 30+ students.
- I need specialist equipment at home to deliver an online lecture.

Teaching Methods And Classroom Environment

- It will be easy to run an online class by myself
- It is not possible to nurture a friendly and cosy classroom environment online.
- Practical exercises must be scrapped as they do not work at an online environment
- Students will not be able to engage and interact in the same way they can in f2f classes.
- It is not possible to provide adequate/individual support to the students.

Planning Ahead

- Online classrooms need not have a cap on the number of attendees (and therefore can be more profitable).
- I can just continue to zoom in and teach.
- Online teaching takes away the lecturer's drive to teach.

Fig. 1 Misconception categories

When teaching is conducted online, it is imperative that the hosting institution provides adequate IT infrastructure for the teaching session to run smoothly. The bare minimum requirements are a stable and fast internet connection and an easily accessible network drive providing shared access to an online hosting platform that meets the relevant pedagogical requirements of the taught content. Teaching staff working within larger institutions may have higher likelihood of access to all these amenities.

Stable internet connection onsite is nearly guaranteed in most tertiary institutions nowadays, but it is less certain when working remotely. Just as one would not expect a lecturer to walk out of a packed lecture theatre in the middle of a lesson, it is embarrassing, awkward and disruptive for either the lecturer or students to be cut off/drop out of an online session in mid-sentence as a result of internet connectivity failure. Similar arguments hold regarding easily accessible and robust shared storage drives (where slide presentations and other course materials are likely to be stored), the connection to which should also not cut out/have access denied. As inevitable as such issues may be, they should not be a regular occurrence simply as a result of poor internet connectivity. Therefore, a centralised, institution-based approach to assist staff with the necessary upgrades should be in place to facilitate the smooth delivery of online teaching sessions.

At UCL, an annual license to Zoom (a cloud-based video conferencing service) was purchased in late summer of 2020 [10]. This meant that UCL staff and students were able to access all of Zoom's services going into the new academic year, 2021–2022. Nonetheless, we explored other video conferencing platforms that enable online course delivery between spring and summer of 2020 at CASC: Microsoft Teams [8] and Moodle-based Black Board Collaborate [11].

These two alternative platforms did not prove to be sufficient for our teaching due to the: requirement for institution-based log-in details, restrictions on the number of simultaneous live cameras feeds, chat and raise-hand functionality, video quality and sound transmission and overall speed capacity with a classroom of 30+ students. In these respects, Zoom proved to be the better alternative, eliminating the barriers mentioned above and providing a new home for online CASC courses.

In most instances, our participants' IT setups proved to be sufficient for attendance with few occurrences of people dropping out but soon enough re-joining the online class. This may not be too surprising considering that participants' needs are limited to (simply) watching a live video stream. We do however recommend that, where possible, participants consider a setup with two screens: one for the Zoom transmission and another for a copy of the slides/notes/software window. In instances where this was not possible, participants reflected on it in their feedback, which has subsequently made us emphasise it even more (especially for software based courses) in pre-course emails, when participants are planning their setup and preparing for the course.

In summary, and based on the conditions outlined at the start of this section related to institution driven flexibility, utilising UCL's existing drive cloud/shared-drive service and Zoom license proved sufficient to address the IT needs for the online delivery of CASC courses.

Misconception 2 I need specialist equipment at home to deliver an online lecture.

Of equal importance to the necessary IT infrastructure provided by the hosting institution is the equipment available to teaching staff during online sessions, either at home or at their worksite. These include the need for additional screens, digital pens for live annotations on projected content, sufficiently powerful computers/laptops and an appropriately healthy/ergonomic workstation.

These can be more easily controlled, maintained and updated as necessary onsite; however, a more coordinated and centrally orchestrated approach is required for teaching staff who choose or have to work from home. A workstation desk, display screen equipment (DSE) and healthy working assessment at both settings are imperative for staff's good physical health because prolonged use of computer workstations can lead to body aches, such as neck, shoulder and back problems.

Funding should be made available to staff to purchase equipment that would enable them to closely replicate the f2f teaching environment online. More specifically, a minimum of two screens makes it possible for the teacher to clearly see the content presented to the students, the live camera feeds enabling eye contact with students, the chat box as well as notes or other resources relevant to the taught material. These items represent a setup similar to f2f teaching where paper notes, a second (laptop) screen and clear visibility of all students present are taken for granted.

At CASC, due to availability of funds locally, we were able to purchase these non-specialist items: second screens and digital pens (feedback below) and were sent desk assessment toolkits by UCL. Where funds might not be available, emphasis should be put by the respective teaching staff to the appropriate institutional team for consideration. A relevant quote from a student is:

The presentations worked well online and the use of annotations was fantastic!

3.2   Teaching Methods and Classroom Environment

Misconception 3 It will be easy to run an online class by myself.

Being the sole person responsible for an online class of 30+ participants can be an unrealistic and tiring expectation and one that no member of teaching staff should have to face alone. From simple (to more complicated) technical issues to demanding students and a class interspersed with hands-on exercises, extra support should be always present.

In teaching groups (by definition, formed by two or more individuals), it should be the norm that a class has at least two members of teaching staff always present. As the audience of CASC has grown over the years, so has the team, in terms of staff members, to ensure that training demands are well met and supported. Availability of several staff members for online classes has played a paramount part to our online delivery transition and has been greatly appreciated by our participants too. A relevant quote from a student is:

I very much appreciated having four instructors on the Zoom so that while one was lecturing the other three could field side-questions.

The number of courses offered should be limited when staff numbers are restricted; smaller class sizes can be better managed by single teaching leads so that the learning outcome and the quality of the class environment are not compromised.

Misconception 4 It is not possible to nurture a friendly and cosy classroom environment online.

Being greeted with a panel of black tiles or still, lifeless profile photos of students at the start of an online session is not anyone's ideal start to the day nor the reason one gets into teaching. Classrooms are meant to be vivid, exciting and stimulating places nurturing students' curiosity and desire to learn. Online classrooms will never have the buzz of a happy f2f class, but feedback from our short course participants at CASC has demonstrated that they can get close. Relevant quotes from students are:

I really enjoyed the course — I have only attended the in-person courses before so was not sure how it would work online, but was really good. Very informative and well-organised.

I was anxious that some of the quality would be lost via Zoom — but having been involved in previous F2F courses — it's worked really well! Very happy to sign up for more:)

This is the first time I have done an online lecture and I was really impressed (I had been dreading it!)

The next two misconceptions further expand on this one (number 4) and on the tools that we used to nurture a friendly class environment.

Misconception 5 Practical exercises must be scrapped as they do not work at an online environment.

Practical exercises on statistical concepts (and beyond) should not be scrapped when re-designing the online delivery of a previously f2f delivered module. The latter would normally feature regular pauses for practicals and these should play an integral part in online lectures too. Pauses during an online lecture may feel awkward and unnecessary, especially as most sessions are (pre-) recorded, meaning students can reply, rewind and forward content as they wish at a later stage.

However, regular pauses give students time and space to at least gather their thoughts in synchronous teaching [12], and also to find the courage to turn on their microphone for questions and to put their new knowledge and skills to test by trying to solve any proposed exercise. The latter might take the form of hands-on practice with statistical tasks on SPSS, for example, pen and paperbased calculations of mathematical nature, discussion on study design and results of existing literature and more. These types of practicals are certainly utilised by CASC on every single short course delivered. Depending on the time available and the class size, these are conducted either in the main Zoom class, where participants are given a 5 or 10 min interval to work by themselves, while the lecturer remains silent, or within breakout groups of 4–6 participants and a lead from the teaching team. Therefore, practical exercises should be welcome and embraced.

Misconception 6 Students will not be able to engage and interact in the same way they can in f2f classes.

During online lectures, the audience (or some, often large, portion of it) may be invisible to the teacher (with no live camera feeds). Therefore, it may be impossible for the teacher to gauge students' engagement level, which would otherwise have been apparent via body language cues, peer-discussion and small group activities.

All the above was overcome at CASC with the following steps:

- Transparency about the importance of students putting their cameras on right at the beginning of each course.
- Regular pauses that give the students time and space to gather thoughts and formulate the right questions in their minds.

- Online polling to gauge understanding via instantaneous feedback from students.
- Availability of additional teaching staff to oversee students' work and engagement in breakout rooms but also for 1-2-1 support where urgent help is needed.
- Appropriate use (but certainly not overuse) of media resources that will help compartmentalise the lecture to a cycle of statistical theory, practical exercises, media use/fun facts, break-our rooms, etc.

Another relevant quote from a student is:

Very well delivered … as an online course, and great that breakout rooms with other facilitators were also available for 1:1 help

Misconception 7 It is not possible to provide adequate/individual support to the students.

Being able to answer questions at a group or individual level is paramount for student satisfaction [3]. Some view statistics as an obstacle to their professional career and progression [1] and one of CASC's top educational objectives is to provide a safe space for students to pose their questions and for us to answer them.

Extra support, as discussed earlier, plays an important role in achieving this goal. The extra member(s) of teaching staff helps by keeping a close eye to the chat function of the online session and addresses questions as necessary or draws the attention of the main speaker to specific questions that would benefit the whole class.

Additionally, breakout rooms provide the required space for one-to-one chats between a student and a teacher or can be extended to a small group of students with the same question. They can also act as networking and socialising spaces if there is such appetite amongst the audience.

3.3   Planning Ahead

Misconception 8 Online classrooms need not have a cap on the number of attendees (and therefore can be more profitable).

Online classes defy the need for a maximum class size as they are not restricted to the physical size of a classroom nor the number of chairs/tables/students it can fit. Should maximum caps of student numbers at online lectures be discarded?

At CASC, we have found this approach to be detrimental to the quality of our teaching and the student experience due to the fact that larger audiences subsequently translate to a potentially larger number of technical issues/difficulties and student questions too. With the same amount of time dedicated to the online lecture, there is simply not enough time to offer the maximum pedagogical experience to students if the size of the audience is substantially increased. As a result, at CASC, we did not increase class sizes more than approximately 5%, allowing only 1 or 2 extra spaces, which we thought was manageable within our resources.

Misconception 9 I can just continue to Zoom in and teach.

It is not often the case that the press of a single button can cause so much disruption in one's screen. But in Zoom, this is a reality. The 'Share screen' button can cause such a significant change to one's screen layout altering the position of windows, slides, live camera feeds and chat box; therefore, only practice makes perfect.

Trial runs with one's own self or colleagues are invaluable and their importance should not be underestimated as they could save one from a lot of headaches. Therefore, our experience tells us that it is never as simple as zooming in and teaching.

Misconception 10 Online teaching takes away the lecturer's drive to teach.

The overnight emergence of online statistics teaching has created a new educational era. After a lot of practice, hundreds of hours of actual teaching in a short period of time, many errors but successes too and a lot of learning along the way, the new reality is fascinating: a much wider audience can be reached and be taught the topic that is loved by so many of us at Burwalls: statistics.

So long as our passion for statistics does not diminish and we are willing to adapt to new ways of delivery, request that our institutions support us in this journey and continue to put students' educational gains to the forefront of our pedagogical objectives, online statistics teaching should not be seen as a barrier.

4　Conclusions

The summary of our experience as described in this chapter and of our realisations and proposed ideas have been driven exclusively by teaching statistics online from March 2020 to September 2022. Teaching staff from other disciplines might find that some common themes are emerging with regards to online teaching in general.

Our experience agrees with findings from Krauss et al. [13] whose results contradict a widespread misconception that students were not adequately engaged with online learning during the 2020–2021 academic year. Similarly, Guevara et al. [14] burst three common myths surrounding online training (about design transition, empathetic approach for learners and socially isolated online environments) via evidence-based case studies. Our call for change in organisation cultures for more effective use of available technologies and resources agrees with Sandras et al. [15] in their presentation of 12 tips for rapid migration to online learning.

It would be naive not to acknowledge scenarios where the myths burst by CASC's recent experiences remain reality for some. Availability of funds, suitable software and hardware solutions, technical resources and cooperation of the target audience are some of the aspects of online teaching that lay beyond the control of those delivering the teaching. In those instances, it is strongly recommended that an institution-wide conversation is initiated to improve the online teaching culture amongst all relevant departments, such as IT and organisational development for training teaching staff.

Face-to-face statistics classes have their own challenges, which teaching staff have simply been very accustomed to over the years — for example, students not being able to locate rooms, transport being disrupted, IT failures such as projectors overheating and computers needing updates. Online statistics teaching has unique challenges and benefits too.

CASC's strategy for lively online interactions has been very well received by short course participants. Therefore, at CASC, we plan to further develop our model to provide high-quality online statistics training that evolves with student feedback and training needs. Students get the same teaching experience with regards to lecturer attention/input, are well supported by assistants where necessary and have access to videos post-course delivery to recap any areas of confusion.

For the majority of our target audience, it seems that they may never wish to return to f2f delivery of short courses; the emphasis being on learning a skill of immediate need rather than forging networks with other course participants. We have therefore learnt to adapt to the new 'normal' of short courses and this has benefits for our audience.

Hybrid short courses, where the f2f lecture is concurrently transmitted live to an online (Zoom) classroom [16], are a very appealing format for short course providers going forward; they certainly are for CASC. The requirements of two groups of participants will be satisfied: participants whose circumstances prohibit them from travelling to the classroom of the f2f class, and also those that seek exclusively f2f training courses due to their own learning preferences. The future of online statistics teaching at CASC is exciting and, most definitely, hybrid.

References

1. Onwuegbuzie AJ, Wilson VA. Statistics anxiety: nature, etiology, antecedents, effects, and treatments--a comprehensive review of the literature. Teach High Educ. 2003;8(2):195–209.
2. Smith AM. Continuing education and short courses. Palliat Med. 1996;10(2):105–11.
3. Moon J. Short courses and workshops: improving the impact of learning, teaching and professional development. London: Routledge; 2014.
4. University College London. Great Ormond Street Institute of Child Health Centre for Applied Statistics Courses. London: University College London; n.d.. www.ucl.ac.uk/stats-courses.
5. Birch JB. Ten suggestions for effectively teaching short courses to heterogeneous groups. Am Stat. 1995;49(2):190–5.
6. Bradstreet TE. Teaching introductory statistics courses so that nonstatisticians experience statistical reasoning. Am Stat. 1996;50(1):69–78.
7. Lowenthal PR, York CS, Richardson JC. Online learning: common misconceptions, benefits and challenges. Hauppauge, NY: Nova Science Publishers; 2014.
8. Microsoft Teams. Microsoft Business Communication Platform; 2021. https://www.microsoft.com/en-gb/microsoft-teams/group-chat-software.
9. Parkes RS, Barrs VR. Interaction identified as both a challenge and a benefit in a rapid switch to online teaching during the COVID-19 pandemic. J Vet Med Educ. 2021;48(6):629–35.
10. Zoom Video Communications Inc. 2021. https://zoom.us.
11. Black Board Collaborate. Virtual classroom solution. 2020. https://www.blackboard.com.

12. Stuart J, O'Donnell AW, Scott R, O'Donnell K, Lund R, Barber B. Asynchronous and synchronous remote teaching and academic outcomes during COVID-19. Dist Educ. 2022;16:1–8.

13. Krauss NE, Divino JN, Crivello JF, Redden JM. Reflections on a plague academic year: a multiyear comparison of student engagement, investment, and performance in undergraduate physiology courses during the COVID-19 pandemic. FASEB J. 2022 May;36:1.

14. Guevara K, Fattah L, Ritt-Olson A, Yin PL, Litman L, Farouk SS, O'Rourke R, Mayer RE. Busting myths in online education: faculty examples from the field. Journal of Clinical and Translational. Science. 2021;5(1):e149.

15. Sandars J, Correia R, Dankbaar M, de Jong P, Goh PS, Hege I, Masters K, Oh SY, Patel R, Premkumar K, Webb A. Twelve tips for rapidly migrating to online learning during the COVID-19 pandemic. MedEdPublish. 2020;9(82):82.

16. Hollenbeck B, Shi Q. Developing effective and sustainable distance education programs and courses. Int J Inform Educ Technol. 2021;11(2):102–6.

Authentic Project-Based Assessment Using the Islands: Instructor's View

Mary Kynn and Nicole B. Reinke

1 Introduction

Learning how to 'read' and 'do' statistics is believed to be challenging for many students [1, 2] including those in the health disciplines. This challenge is more pronounced for students who did not study mathematics above a basic secondary school level [3], or are returning to study as a mature student. Additionally, many students find that there is a disconnect between introductory techniques they are taught in a statistics class, and the ability to understand the variety of techniques, terminology, and presentations of statistics that are used in published research. For postgraduate students the challenge is in rapidly applying statistical knowledge to the rest of their program where they are likely to conduct literature reviews, and design and complete their own research study. One solution is to incorporate authentic projects into the statistics curriculum; where students design a small study, collect and analyse their data, and then present the results in a scientific format. This approach is not new; see, for example, Mackisack [4] who found her inspiration to use project-based learning to teach statistics in the earlier report of student projects by Hunter [5]. The use of project-based learning has been standard in some universities for more than 30 years (the first author learned in this manner as an undergraduate student). However, using real life investigations limits the scope of research questions to those which are ethical, small scale, and practicably possible within a short time frame. These constraints can make it especially challenging for health students who are likely to be interested in health and human related research questions, potentially of an ethically 'risky' nature for an inexperienced researcher. This paper describes teaching experiences from the last 5 years of designing and delivering

M. Kynn (✉)
Faculty of Science and Engineering, Curtin University, Bentley, WA, Australia
e-mail: mary.kynn@curtin.edu.au

N. B. Reinke
University of the Sunshine Coast, Sippy Downs, QLD, Australia

unique project-based assessment using a virtual, web-based population in a simulation called the *Islands* [6] in courses (subjects) in quantitative research methods, statistics, and epidemiology.

1.1   Project-Based Learning and Assessment

Project-based learning is a way of instruction that engages the learner in authentic problems [7]. The early use of project-based learning took place in school settings where it has been reported to have positive effects on students' motivation, engagement, and self-efficacy [8]. When used in university education it allows students to engage in real-world activities that are similar to those experienced in professional practice. As such, project-based learning has been reported to promote deep learning and to allow students to learn by practising the skills they will need in future study and for career success [8].

Krajcic and Blumenfeld [9] describe learning environments using a project-based approach as having five key features.

1. Start with a problem or driving question.
2. Exploration of the problem by engaging in authentic, situated inquiry.
3. Collaboration between students, teachers and community members to find solutions.
4. Engagement in advanced activities, supported by scaffolding.
5. Production of products/solutions to the problem.

A 'project' in statistical and epidemiological education typically involves developing one or more research questions, custom designing methodology for data collection (and carrying this out), data analysis and visualisation, interpretation of the analysis, and communication of the findings relative to the research question. Finally, the students have the opportunity to reflect on the limitations of the design and the ability of the design to fundamentally answer the question of interest. Project-based assessment promotes students' feelings of ownership over various aspects of their learning and builds confidence in their own abilities to read and understand research. In contrast, if a hypothesis and data set are provided, although the student decides on and conducts the analysis, their interpretation of the data relative to the research question is limited by the amount of context provided. Typically, conclusions are reduced to an 'accept' or 'reject' statement and the student has little insight into how the data are contributing to the overall research goals. This disconnect is exacerbated when the hypothesis has been retrofitted to the data in order to meet specific learning objectives, essentially contravening the scientific method. With project-based assessment students tackle the scientific method in the correct order (observation, hypothesis, design, data, analysis, conclusions, more observations) and take 'ownership' of each step. They can engage more deeply with the process and develop crucial skills in being able to critique published research by critiquing their own.

The difficulty with health research is that it can be impractical and unethical for students to conduct small projects themselves in a short time frame. A virtual population solves this issue and facilitates rich learning experiences which may not be possible in real life.

1.2   A Virtual Population

The *Islands* is a web-based interface developed by Bulmer [6]. It originally consisted of a single island ('the *Island*') and was expanded in 2015 to include a collection of islands, increased infrastructure and social measures, to become 'the *Islands*' (Fig. 1). Further details of the *Island*, the *Islands,* and about how to gain (free) access are detailed in Bulmer and Haladyn [6, 11] (registration at https://islands. smp.uq.edu.au/register/).

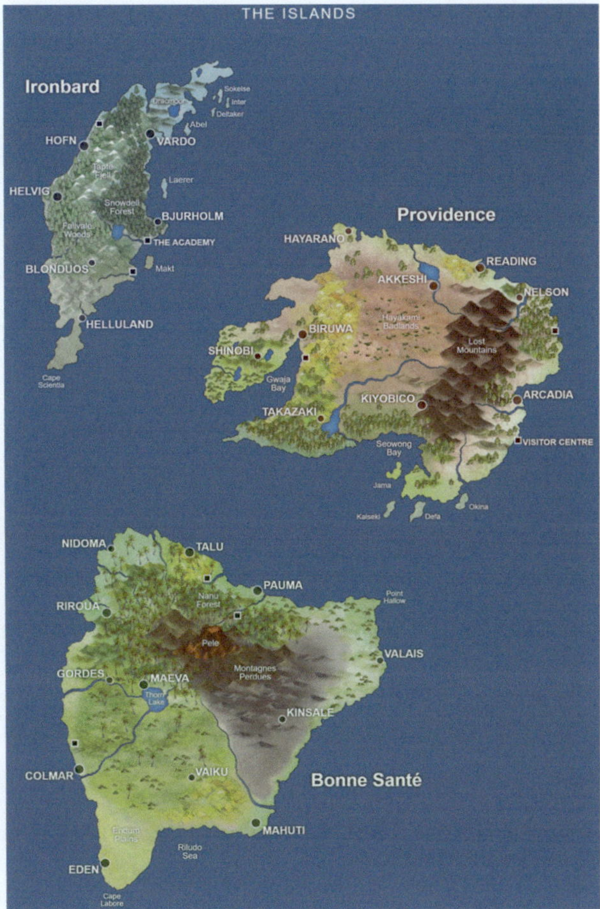

Fig. 1 The interface to begin exploring the *Islands* [10]. Using the interface, locations can be clicked to see the further details

The *Islands* are a population of virtual human subjects who can be accessed for both quantitative and qualitative data collection [11], although in our context we were focused on quantitative data collection. The 'history' of the *Islands* is that of a shipwreck more than three centuries ago which stranded a small group of people in this remote location. Over time they formed thriving communities and spread out across the neighbouring islands; however, they remained isolated from the rest of civilisation and as such are an under-studied population. *Islanders* have built a familiar infrastructure, with each village containing a school, a clinic, a hall of records, and a bureau of information, and the larger towns have hospitals and universities. There is also a museum and a growing number of police stations.

Residents of the *Islands* can be found in many ways: in their houses, through their school or employment records, and through records of births, deaths, or marriages (Fig. 2). They can also be found through their recreation groups and medical and police records.

Once a person has been 'found' or 'sampled', a researcher must ask for consent to interact with them. If consent is given, this indicates that the *Islander* is willing to take many tests and perform a variety of tasks (medical, psychological, physical, etc.), answer questions, have their records interrogated, and even 'chat'. It is

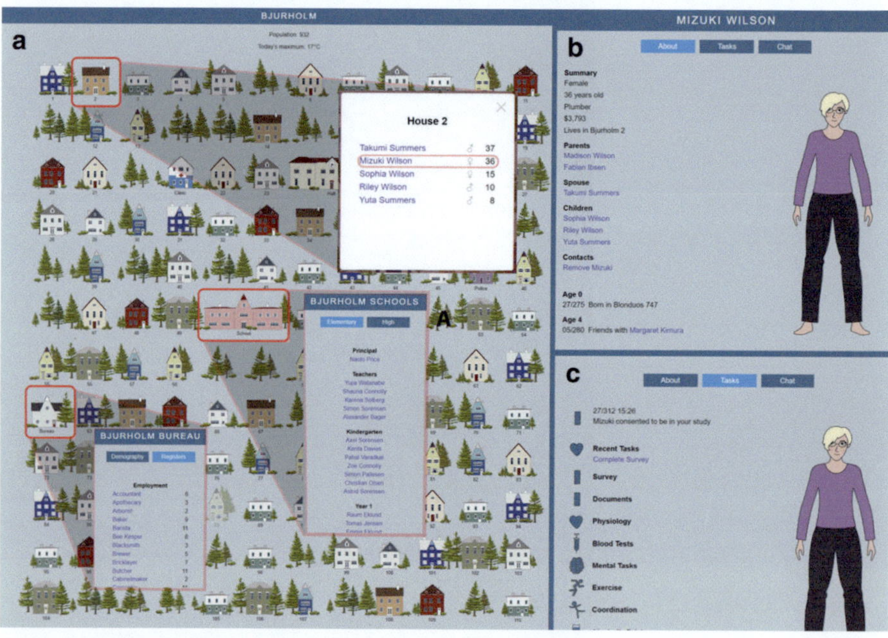

Fig. 2 Composite screen capture from the *Islands* [10] showing in (**a**) the houses and some of municipal buildings and how they link to lists of people; (**b**) the 'About' page for one of the Islands residents; and (**c**) the 'Tasks' tab which allows the user to ask for consent and provides lists of tasks the Islander can undertake

possible to see their life history as well as the history of any interactions they have had with the researcher. However, any interactions with other researchers are invisible.

Many aspects of the *Islands* were deliberately included to mimic real research: not every house is occupied; not everyone will consent; a young child cannot answer questions; there is a time 'cost' to any interaction; occasionally answers to questions don't exactly match records (for example, an *Islander* may state they are older or younger than they truly are); occasionally they can't answer a question; and they may move away or die part-way through a study. In addition to this, it is not possible to know the outcome of a study before it begins, even for the creator of the *Islands*.

2   Context and Design

Projects using the *Islands* were incorporated into a total of three semester-long courses at the University of the Sunshine Coast, Australia, all having statistics as the primary focus, but also included elements of design, ethics, and epidemiology. It was first introduced in a third-year undergraduate course, Epidemiology and Biostatistics, in 2013 (80–150 students per cohort). This is a required course in more than five health-related bachelor's programs. The details of this were described by Dunn et al. [12]. It was then integrated into two postgraduate courses. Quantitative Research Methods is a required course for all first year Master of Health Promotion (MHP) students (20–100 students per cohort), optional for Master of Clinical Nursing students, and available for any postgraduate students across the university as an elective or as a replacement (alternative semester) statistics course. Epidemiology and Biostatistics (postgraduate version) is a required second-year course for MHP students (20–30 students per cohort). The postgraduate courses were delivered (including marking) by a single coordinator, with an additional staff member providing moderation. The undergraduate course coordinator had the support of a tutor most years. The experience of using the *Islands* in one undergraduate and two postgraduate courses, from 2015 to 2020, is presented in this chapter.

The Master of Health Promotion program attracted both an international cohort who primarily had a background in a clinical area (such as nursing), and a smaller number of online domestic students who were often returning to study after a significant break. Although in theory all students should have encountered basic statistics in their undergraduate degrees, for many, this foundational knowledge was missing or forgotten. Almost no students had previous experience using SPSS. Many students were not familiar with the learning management system (Blackboard), and some students had not used basic Office applications (Word, Excel, PowerPoint). Although all students had completed an undergraduate degree, this may have been examination based. Thus, many students were still learning how to read and write in an academic setting. In addition, English was a second (or third) language for most of the international students.

When incorporated into Quantitative Research Methods, the *Islands* based assessment replaced an 'own-choice' project-based assessment where students developed a small research question for which they collected and analysed their own data in real life. In Epidemiology and Biostatistics (undergraduate and post-graduate versions) the *Islands* based assessment replaced a data analysis assignment where students were provided with an epidemiological data set and research question. In each case the *Islands* based project was a major assessment piece (up to 40% of final grade) which was due at the end of semester and supported by earlier low-weighted assessment tasks throughout the semester.

2.1   Class (Pre-project) Activities

Topics in Quantitative Research Methods included introductions to data collection, descriptive and inferential statistics (including t-tests, ANOVA, chi-squared tests, and linear regression). Topics in Epidemiology and Biostatistics included study designs, ethics, sampling, revision of parametric techniques, non-parametric techniques, and logistic regression. The *Islands* were introduced in all courses before the project work by using it in workshop activities. Note that not all of the following activities were used in all years, or in all courses.

1. *Summary statistics and graphs.* Using the *Hall*, death records can be found and copied into statistical software for simple tallying or creation of bar charts. As a class, the causes of death can be compared across towns, with each student investigating a separate town. This activity once unveiled that a town had at one time been the victim of a volcanic eruption with the leading cause of death being related to ash coverage. Through the *Bureau* suitable summary demographic data for a population pyramid can be found for a similar activity.
2. *Sampling frames and sampling techniques.* Students can be given a list of hypothetical research questions for them to propose/design suitable sampling strategies. This makes sampling a very visual/tangible experience and greatly increases the speed and depth of understanding.
3. *Community or prevalence survey.* As a class it is possible to collect data collaboratively onto a Google spreadsheet (Fig. 3). This activity makes for a very thorough review of issues on data collection and cleaning and also creates a large data set relatively quickly which can be refreshed each year on different topics or populations. This data can be used as the basis for power calculations for a later project where suitable.
4. *Writing good research questions.* Students can review the PICO (Population Intervention Comparison Outcome) format for writing good research questions whilst exploring the Islands and developing possible topics for research. Students need to check they can sample the appropriate population, measure appropriate outcomes, and apply appropriate interventions (if needed). If a variable is not available (such as measuring alcohol intake), they can consider alternatives (such as measuring liver size).

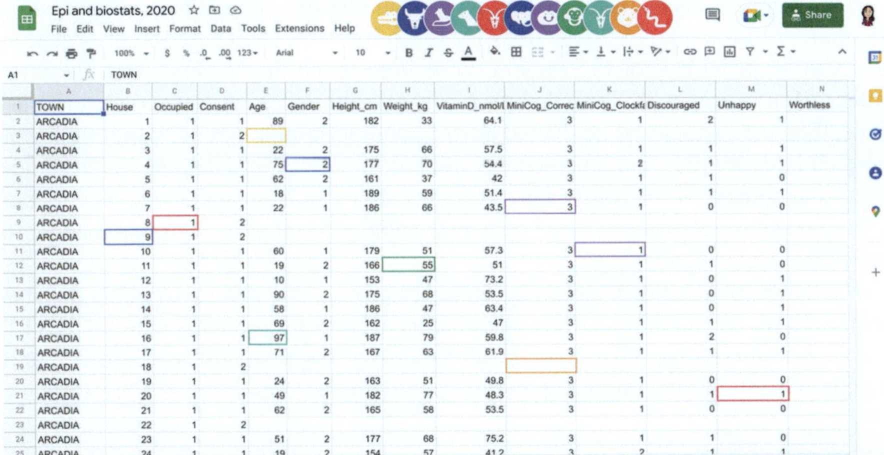

TOWN	House	Occupied	Consent	Age	Gender	Height_cm	Weight_kg	VitaminD_nmol/l	MiniCog_Correc	MiniCog_Clockf	Discouraged	Unhappy	Worthless
ARCADIA	1	1	1	89	2	182	33	64.1	3	1	2	1	
ARCADIA	2	1	2										
ARCADIA	3	1	1	22	2	175	66	57.5	3	1	1	1	
ARCADIA	4	1	1	75	2	177	70	54.4	3	2	1	1	
ARCADIA	5	1	1	62	2	161	37	42	3	1	1	0	
ARCADIA	6	1	1	18	1	189	59	51.4	3	1	1	1	
ARCADIA	7	1	1	22	1	186	66	43.5	3	1	0	0	
ARCADIA	8	1	2										
ARCADIA	9	1	2										
ARCADIA	10	1	1	60	1	179	51	57.3	3	1	0	0	
ARCADIA	11	1	1	19	2	166	55	51	3	1	1	0	
ARCADIA	12	1	1	10	1	153	47	73.2	3	1	0	1	
ARCADIA	13	1	1	90	2	175	68	53.5	3	1	0	1	
ARCADIA	14	1	1	58	1	186	47	63.4	3	1	0	1	
ARCADIA	15	1	1	69	2	162	25	47	3	1	1	1	
ARCADIA	16	1	1	97	1	187	79	59.8	3	1	2	0	
ARCADIA	17	1	1	71	2	167	63	61.9	3	1	1	1	
ARCADIA	18	1	2										
ARCADIA	19	1	1	24	2	163	51	49.8	3	1	0	0	
ARCADIA	20	1	1	49	1	182	77	48.3	3	1	1	1	
ARCADIA	21	1	1	62	2	165	58	53.5	3	1	0	0	
ARCADIA	22	1	2										
ARCADIA	23	1	1	51	2	177	68	75.2	3	1	1	0	
ARCADIA	24	1	1	19	2	154	57	41.2	3	2	1	1	

Fig. 3 Screen capture of collaboratively collecting data on a Google spreadsheet in class, which included some online students who could see in real time exactly what the class was doing and participate effectively

2.2 Project Activities

Following training of how to use the Islands in workshops, the project was largely completed outside of class time, although class time was available for progress discussions and quick consultations. Students were able to develop their own specific research questions within the guidelines given. In Quantitative Research Methods students were required to use a cross-sectional design, have a single continuous outcome, a primary explanatory variable, and an additional three or four explanatory variables. The final analysis presented by the student would fall in the family of general linear models. This was an individual task although students were permitted to choose similar topics. In Epidemiology and Biostatistics students were required to design a randomised controlled trial, using the class prevalence survey data for power calculations. The outcome in this case was required to be binary, and they were encouraged to have several explanatory variables in addition to the intervention variable. The final analysis was a logistic regression. This was a group task for students in both versions of the course, although some students chose to work individually.

In most applied statistics courses, at the time that students are designing the study, they will not be able to understand the analysis they will later need to perform on the data. It is therefore important to firmly direct the students as to the nature of the types of variables they should be collecting. In these courses students were directed to whether the study should be cross-sectional or experimental; however, it would be possible for students to use other designs, such as cohort and case-control.

In addition, many steps in completing a small research project may not necessarily align with the learning objectives for the subject. In each case the information needs to be provided for the student, or at least made into a very quick decision process. For example, if the students are not required to demonstrate power calculations, then they should be instructed as to a suitable sample size; if sampling is not covered in depth they can be guided to suitable strategies; if a detailed understanding of the context of the study is not required, they can be given advice on variable selection and if necessary advised on a reasonable amount of background reading to perform.

It is advisable to either have a timeline for project completion to guide the students or have them complete this planning as an activity (perhaps with a GANTT chart). Where possible, it is recommended that key stages should be scaffolded into weekly workshops and aligned with the teaching materials.

3 Discussion

3.1 Strengths of *Islands* Based Project

The greatest strength of *Islands* based projects is the authenticity of the experience, for the relatively small time-investment. As mentioned previously, there are situations that students may encounter when collecting data, such as empty houses, non-consent, non-availability, and mortality during the study, where students will need to decide how and if to record this information. In addition, they will also need to navigate other genuine research difficulties. These include occasions when answers to questions do not make any sense, or if all participants answer a question in a similar way. Students may get exactly the answer they expect with a satisfying level of significance, but more often, they will not. In that case, they will need to reflect on the design of the study and decide if the 'absence of evidence' is 'evidence of absence'. Students may also be able to reflect on strengths of their design even if the results were unexpected. Usually, students will have ideas about how the study could be improved for future research, aside from the idea of increasing sample size which is a staple discussion point for most student projects. Occasionally students demonstrate a depth of ethical understanding which is also a critical skill for novice researchers. For example, if there is evidence of an intervention actually causing harm; students may reflect that it is not ethical to do further study without conducting more background research first. An additional strength of using the *Islands*-based projects is the positive student experiences, for example, Dunn et al. [12] reported favourable student perceptions of the *Island* being a good learning experience, beneficial, and easy to use.

3.2   Considerations for Implementing Project-Based Learning with a Virtual Tool

In project-based learning, the cognitive load for students is usually high at any point in time, but can be especially high for those taking foundation courses in the first semester of university study. Although the *Islands* are relatively simple to use, they are still another obligatory 'thing' the student must grasp. If students are just beginning or are returning to university after a leave of absence, it may be more important to focus on other requirements in the first few weeks of the teaching period. Such requirements may include: catching up on basic mathematics or statistics, learning statistics software; and learning how to study or use the content management system (e.g. Blackboard). For a student at any level of study, it is critical to scaffold the project tasks within the learning materials, as at this stage of the learning process, the students cannot 'see ahead' to the analysis. Thus, the project needs to be carefully guided to make sure the students will end up with usable data. Splitting a project into two parts, for example, a proposal or protocol and then a final report or poster, mimics both the real-world experience of conducting research and allows for intervention early in the process to ensure a 'successful' project is completed.

The traditional design of a data analysis task starts with teaching the content and is followed by summative assessment which involves limited discussion or feedback. Such an approach is likely to be counterproductive as it is not consistent with how an authentic project-based task is conducted. Students will want to discuss their ideas and experiences, and using project-based learning as described above, this desire for engagement can be harnessed for both energetic class discussions and group work. Encouraging a variety of specific research questions within an overall research framework means that students can legitimately collaborate (with or without groupwork) whilst still maintaining integrity—indeed, robust discussion with peers is what we would strive to achieve in any research situation. The scope and variety of research questions used across the class will determine if there need to be any boundaries put on discussions. However, even if a similar topic is undertaken by every student, the nature of using the *Islands* is that they will each obtain unique data which limits misconduct, although does not preclude it altogether.

As with any student research project there needs to be an allowance for 'supervision' incorporated into the teaching and learning time. Many students will need to discuss the project with a tutor and it is advisable to have a scheduled time available each week, or in critical phases, to hold short individual or small group consultations (2–5 min).

Finally, a marking guide or rubric should be specifically developed for the project to help align the assessment activities with the intended learning outcomes. This is most effective as a standards-based rubric [13] rather than a more traditional 'answers'-based marking scheme.

3.3   Will this Suit Me as a Lecturer?

As a course coordinator/lecturer/tutor there can be learning curve to both loosen the reins on the assessment but also being more involved in an advisory capacity. It is important for students to retain control over their work, but they also require small amounts of ongoing attention (such as at the end of workshops) to ensure they don't go off track or get lost in the proverbial woods. This is similar to meeting with a research-based postgraduate student except that the questions and problems are usually much more manageable. This style of assessment is unlikely to suit those instructors who prefer a didactic (sage on the stage) style of teaching and is equally unsuited to those who prefer students who are self-directed (guide on the side). Instead, the third option of the 'meddler in the middle' is the most appropriate space for the lecturer/tutor to occupy: usefully ignorant, seriously playful, epistemologically agile, and low threat/high challenge [14].

4   Conclusions

The *Islands* are an innovative virtual population who are easy to access and research. Using the *Islands* for authentic project-based assessment in (applied) statistics courses requires some advance planning to integrate it appropriately into the learning materials. However, the benefits to the students of engaging in such a rich and deep learning experience, and to the lecturer of being part of an authentic learning practice, greatly outweigh the costs.

References

1. Dunn DS, Smith RA, Beins BC. Preface. In: Dunn DS, Smith RA, Beins BC, editors. Best practices for teaching statistics and research methods in the behavioral sciences. New York: Routledge; 2012. p. xv–xvii.
2. Hannigan A, Hegarty AC, McGrath D. Attitudes towards statistics of graduate entry medical students: the role of prior learning experiences. BMC Med Educ. 2014;14(1):70. https://doi. org/10.1186/1472-6920-14-70.
3. Donnison S, Dunn PK, Cole R, Bulmer M, Roiko AH, Muller F. Enhancing curriculum in epidemiology and biostatistics through simulation-based learning. Int Res Educ. 2016;4(1):11–26. https://doi.org/10.5296/ire.v4i1.8064.
4. Mackisack M. What is the use of experiments conducted by statistics students? J Stat Educ. 1994;2:1. https://doi.org/10.1080/10691898.1994.11910461.
5. Hunter WG. Some ideas about teaching design of experiments with 2^5 examples of experiments conducted by students. Am Stat. 1976;31(1):12–7. https://doi.org/10.1080/0003130 5.1977.10479185.
6. Bulmer M, Haladyn JK. Life on an Island: a simulated population to support student projects in statistics. Technol Innov Stat Educ. 2011;5(1):187. https://doi.org/10.5070/t551000187.
7. Blumenfeld PC, Soloway E, Marx RW, Krajcik JS, Guzdial M, Palincsar A. Motivating project-based learning: sustaining the doing, supporting the learning. Educ Psychol. 1991;26(3-4):369–98. https://doi.org/10.1080/00461520.1991.9653139.

8. Larmer J, Mergendoller J, Boss S. Setting the standard for project based learning: a proved approach to rigorous classroom instruction. Alexandria: ACSD; 2015.

9. Krajcik JS, Blumenfeld PC. Project-based learning. In: Sawyer RK, editor. The Cambridge handbook of the learning sciences. New York: Cambridge University Press; 2006. p. 317–33.

10. Bulmer M. The Islands. https://islands.smp.uq.edu.au/login.php?req=is-land@maths.uq.edu.au.

11. Bulmer M, Haladyn JK. Virtual environments for teaching qualitative research methods. In: Sorto MA, White A, Guyot L, editors. Looking back, looking forward proceedings of the tenth international conference on teaching statistics (ICOTS10 July). Kyoto, Japan, 2018, p. 1–6.

12. Dunn PK, Donnison S, Cole R, Bulmer M. Using a virtual population to authentically teach epidemiology and biostatistics. Int J Math Educ Sci Technol. 2017;48(2):185–201. https://doi.org/10.1080/0020739x.2016.1228015.

13. Hendry GD, Armstrong S, Bromberger N. Implementing stands-based assessment effectively: incorporating discussion of exemplars into classroom teaching. Assess Eval High Educ. 2012;37(2):149–61. https://doi.org/10.1080/02602938.2010.515014.

14. McWilliam EL. Teaching for creativity: from sage to guide to meddler. Asia Pac J Educ. 2009;29(3):281–93. https://doi.org/10.1080/02188790903092787.

An Interactive Application Demonstrating Frequentist and Bayesian Inferential Frameworks

Mintu Nath

1 Introduction

The frequentist and Bayesian methods provide statistical frameworks to infer the population parameters and associated uncertainties, but they have distinct approaches to achieving this goal. Frequentist methods are taught at beginner to advanced levels of all statistical courses, but Bayesian methods have limited offerings [1]. With the advances in computational power, wider usage of Bayesian tools in different scientific fields, and the current debate on statistical significance from the frequentist viewpoint [2, 3], statistical thinking in the Bayesian paradigm is recognised as essential in the statistics curriculum [1, 4].

Resources on Bayesian education are scanty and often too theoretical to implement at the beginner level, with some attempts to introduce the concept using interactive applications [5, 6]. We describe a simple, user-friendly, interactive application in the R Shiny environment to communicate the key ideas in both inferential frameworks [7]. The application employs a simple real-life example and creates a multitude of scenarios with the help of instantaneous simulations, variations of inputs, a dynamic graphical interface, and user interactivity. The objective is to support learners—from beginner to advanced levels—to understand the principles and analytical approaches of both inferential methods. The paper gives an overview of the application intertwined with the underlying theory for supporting educators to integrate the application for different levels of learners.

M. Nath (✉)
Institute of Applied Health Sciences, University of Aberdeen, Aberdeen, UK
e-mail: mintu.nath@abdn.ac.uk

© The Author(s), under exclusive license to Springer Nature Switzerland AG 2023
D. J. J. Farnell, R. Medeiros Mirra (eds.), *Teaching Biostatistics in Medicine and Allied Health Sciences*, https://doi.org/10.1007/978-3-031-26010-0_11

2   Accessing the Application

1. Online access: using a web browser

   The persistent uniform resource locator (PURL) of the application is: https://purl.archive.org/purl/abacus/app_FreqBayes. The URL of the application is given under target. Please click the link to access the application. The access may, however, depend on the server status.
2. Local access: using the R environment

   In future, the application will be integrated with the R package ABACUS [8] (Apps Based Activities for Communicating and Understanding Statistics). It will require a one-time installation of R [9] and the package. No prior knowledge of R is necessary.

   Step 1: Install R from www.R-project.org/
   Step 2: From the R command line, install the package ABACUS:
   install.packages('ABACUS', dependencies=TRUE)
   Step 3: Run the following codes from the command line:
   library(ABACUS)
   shiny_FreqBayes()

3   Structure of the Application

The application has four sections: *Frequentist Simulate, Frequentist Apply, Bayes Simulate,* and *Bayes Apply.* This structure corresponds with the simulation and application-based approaches to teaching Bayesian statistics [10]. The *Simulate* section simulates the data (when the truth is known) to explain the theoretical basis of each method, and the *Apply* section applies the methodology to the experimental dataset (the real scenario when the truth is unknown) as one can conduct in a standard software environment. The four sections represent the four *tabs* on the top of the application.

Users must provide *inputs* on the *left-hand panel* of each section. Inputs represent the necessary parameters to simulate the data, or relevant information to fit a model. Although conceptually, it means providing the input data in the *Apply* section, the application simulates the data for the users that they will generally provide. The *outputs* for each section are presented on the right panel (with additional tabs) when the corresponding *command button* (on the left-hand panel) is pressed. The inputs and outputs of each section are detailed later.

This paper highlights the statistical theory (described under *Theory*) in both frequentist and Bayesian frameworks in the following four sections. The theory can be omitted for beginners. Educators can use the theory to demonstrate the fundamentals and exemplify the similarities and differences between the two inferential methods in the context of methodologies, outcomes, and interpretations.

Educators can adapt the application in a real problem-solving scenario that follows a *binomial distribution*. For illustration purposes, we set the scene to estimate the *prevalence* of flu in a town (population). We record participants' responses about their flu status in the last winter using a simple random sampling approach. The paper describes the estimation of true prevalence in the population for both frequentist and Bayesian methods. Finally, the article highlights specific learning objectives, lesson plans, and activities to integrate the application for beginner, intermediate, and advanced learners.

4   Frequentist: *Simulate*

Figure 1 shows the screenshot of the inputs (A) and outputs (B-E) of the simulation-based approach in the frequentist framework. The paper presents the theoretical basis of frequentist methodologies and relevant statistical tools and terminologies. Many textbooks explain these statistical principles in detail [11].

4.1   Inputs

Binomial Distribution: Probability (θ)
To simulate the data, we assume the truth is known, i.e., the *true population parameter* θ (for our example, true disease prevalence) is known. The parameter θ represents the probability of the *binomial distribution*. The simulation-based approach emphasises the theoretical underpinning of statistical tools to estimate the true prevalence. Select the value of θ ($0 \leq \theta \leq 1$).

Sample Size (n)
We collect data from the population using a standard *sampling* technique; the *sample size* (n); range allowed ($50 \leq n \leq 1000$).

Number of Replications (k)
In a frequentist framework, we formulate the *relative frequency* concept (see the next section for details). The application captures this concept by selecting the number of *replications* (k); range allowed ($10 \leq k \leq 1000$).

Seed Value for the Random Number
Since the application simulates the data, setting the *seed value* would allow sampling the identical dataset given a set of input values.

Click *Simulate* command button to simulate the frequentist framework.

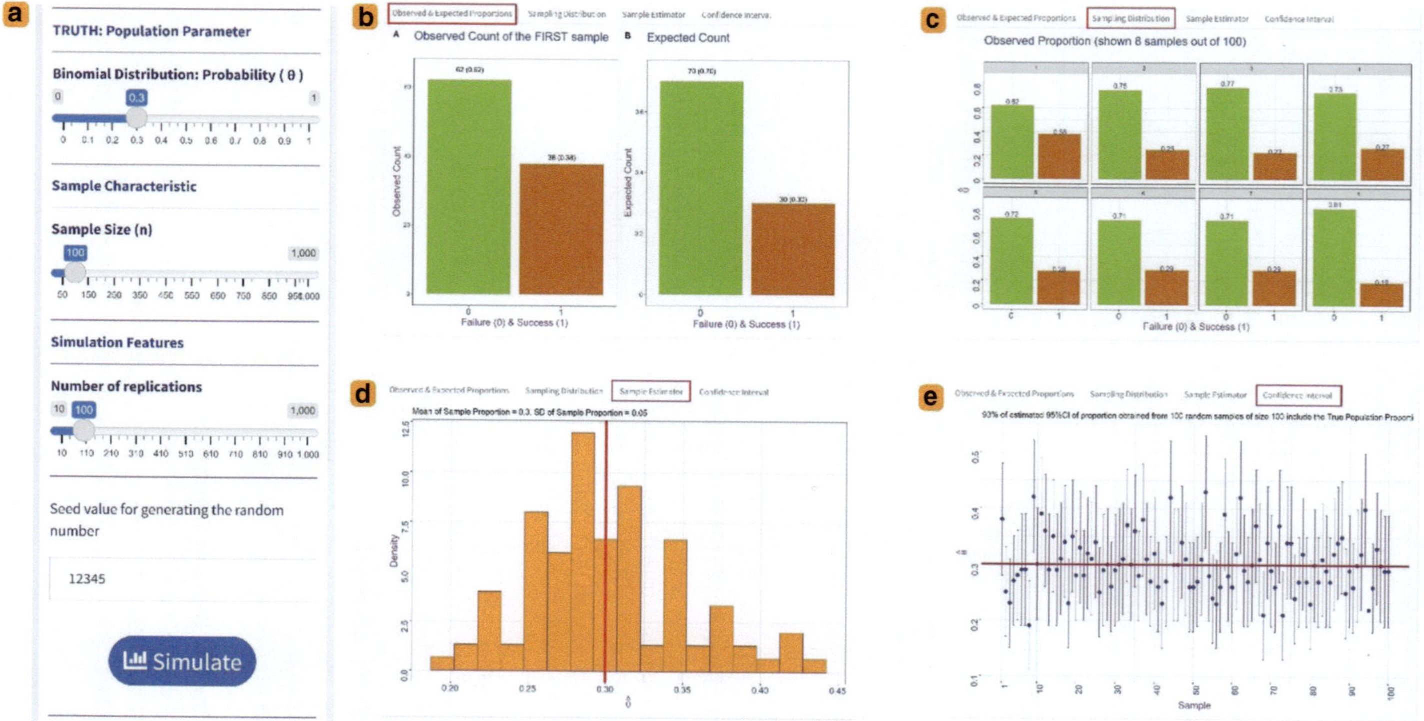

Fig. 1 Screenshot of *Frequentist Simulate* section showing the (**a**) input panel to set simulation parameters; (**b**) observed relative frequencies and expected counts (*Observed and Expected Proportions*); (**c**) sampling distribution of first eight samples (*Sampling Distributions*); (**d**) distribution of sample proportions of all replicates (*Sample Estimator*); (**e**) 95% confidence interval of sample proportions of all replicates (*Confidence Interval*)

4.2   Outputs

Observed and Expected Proportions

Theory:
*The parameter value in a frequentist framework is a **fixed** quantity; it does not have a distribution.*
If the random outcome variable y follows a binomial distribution with parameters n(≥ 1) and θ (0 ≤ θ ≤ 1), then:

$$f\left(y|n,\theta\right) = Bin\left(n,\theta\right) = \binom{n}{y}\theta^{y}\left(1-\theta\right)^{n-y}$$

Mean = nθ; Variance = nθ(1 − θ).

In Fig. 1b, the *observed outcome* of the *first* sample represents a random sample from $f(y|n, \theta)$. The *observed proportion* summarises the *first* sample. This will change each time *Simulate* button is clicked.

Based on the properties of the binomial distribution, the *expected count* represents the *mean* of the *binomial distribution Bin(n, θ)* over all random samples given by *success = nθ* and *failure = n(1 − θ)*.

Sampling Distribution

Theory:
In the frequentist framework, we can repeat the sampling process (or a trial) infinite times under identical conditions. Therefore, the frequentist view of the probability is based on relative frequency, which defines an event's probability as the limit of its relative frequency in an infinite number of trials.

Figure 1c shows the *observed proportion* of the first eight random samples drawn from $f(y|n, \theta)$. Note the *first* sample represents the sample presented in the previous section.

Sample Estimator

Theory:

If we draw random samples of size n from a population many times and calculate the sample estimate $\hat{\theta}$ (i.e., the relative frequency of success for each sample in this case) each time, then the mean of all sample estimates will be equal to the population parameter θ. In this case, the sample estimate forms an 'unbiased' estimate. In statistical language, we can show that the expectation of sample estimate $\hat{\theta}$ is the population parameter θ, i.e., $E\left(\hat{\theta}\right) = \theta$.

For a large enough sample size, if groups of n individuals were sampled repeatedly and truly randomly, it follows from the central limit theorem (CLT) that the distribution of $\hat{\theta} \sim Normal\left(\theta, \sigma_\theta\right)$, where $\sigma_\theta = \sqrt{\theta\left(1-\theta\right)/n}$. The σ_θ is called the standard error of $\hat{\theta}$.

Reasonable conditions for large sample size are: $n\theta \geq 10$; $n(1-\theta) \geq 10$.

Figure 1d demonstrates the *sampling distribution* of the *mean* and *summary* of *k sample proportions*.

Confidence Interval (CI)

Theory:

The interpretation of probability is the long-run relative frequency of a repeatable event.

95% confidence interval (CI) means if the trial is repeated an infinite number of times and 95% CI is calculated each time, then 95% of those CI would include the true parameter value.

The CI is random as it depends on repeated random samples.

For a large sample size, the 95% CI is computed as: $\theta \mp 1.96\sigma_\theta$. After observing the sample, we can estimate $\widehat{\sigma}_\theta = \hat{\theta}\left(1-\hat{\theta}\right)/n$. Therefore the 95% CI is calculated as: $\hat{\theta} \mp 1.96\sqrt{\hat{\theta}\left(1-\hat{\theta}\right)/n}$.

Figure 1e shows the calculated *95% CIs* of *k* samples with the *true population parameter* as a solid horizontal red line.

4.3   Activities

- Change the sample size and discuss the effect of sample size on the mean and standard error of the estimate and the precision based on the width of 95% CI.
- Change the number of replications and explain the mean, standard error, and distribution of the sample proportions.
- Explore other values of θ. What will happen if we consider θ equals to 0 or 1?

5   Frequentist: *Apply*

Figure 2 shows the screenshot of inputs (A) and outputs (B-D) of the *Apply* section in the frequentist framework. For our example, the experimental data constitute the flu status of a random sample of n individuals. We will apply frequentist methodologies to analyse the single instance of experimental data and estimate the prevalence of the disease.

5.1   Inputs

Sample Size (n)
Select the *sample size* (n).

Hypothesised Population Parameter (θ_0)
In the *null hypothesis* (H_0) testing scenario (and H_1 as the *alternative hypothesis*), we hypothesised the population parameter θ as θ_0, i.e., $H_0 : \theta = \theta_0; H_1 : \theta \neq \theta_0$.

Significance Level or Type 1 Error (α)
The *significance level* or the probability of rejecting a true null hypothesis.
   Click **Run NHST** button to apply the frequentist framework on the observed dataset.

5.2   Outputs

Observed Data
Figure 2b presents a grid of all observed data (n) showing the occurrence of the *event* (1, disease state) and *non-event* (0, healthy state). It also presents a bar plot of the summary data.

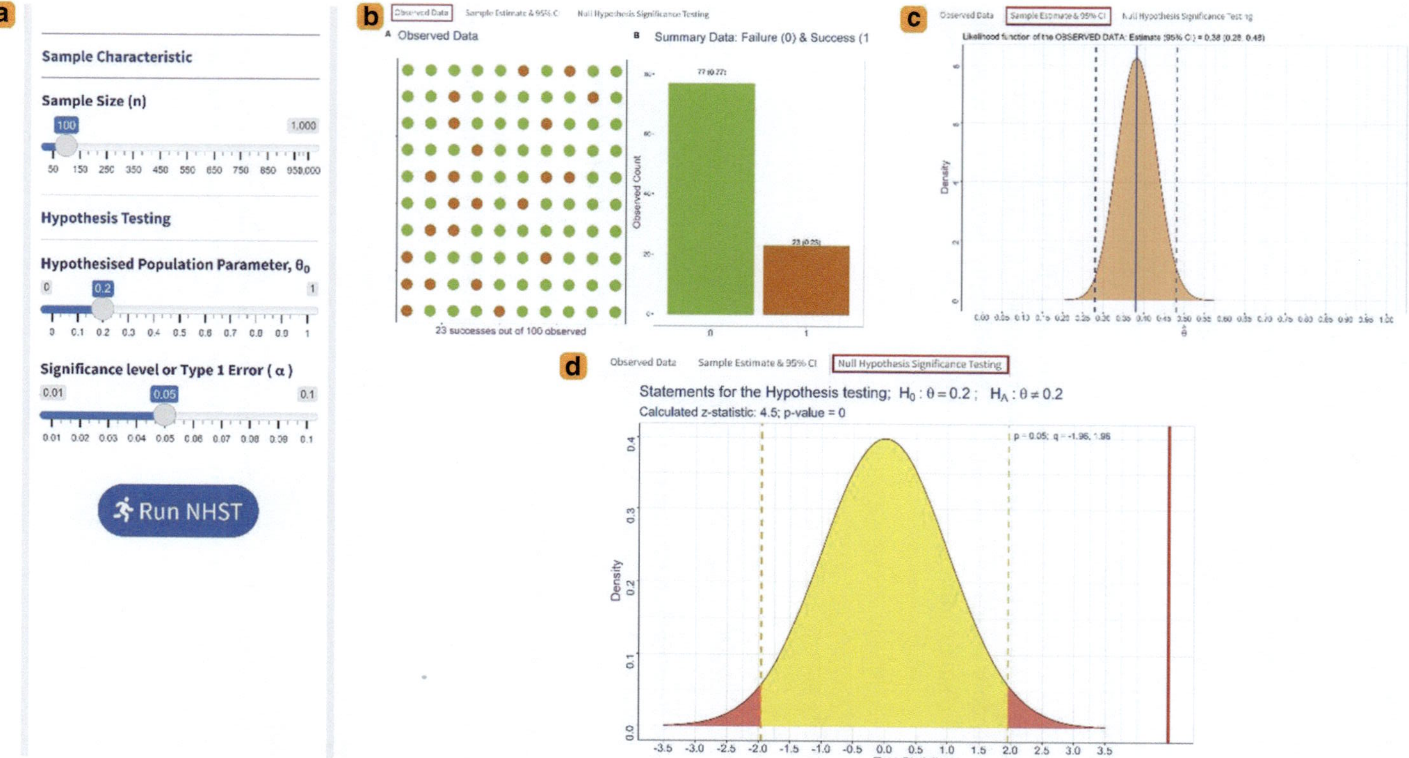

Fig. 2 Screenshot of *Frequentist Apply* section showing (**a**) the input panel to select the sample size, null hypothesis and type 1 error; (**b**) a view of the single instance of the (simulated) observed data and the summary information (*Observed Data*); (**c**) sample estimate with 95% confidence interval (*Sample Estimate and 95% CI*); (**d**) the hypothesis testing statements, distribution of the test statistic under the null hypothesis and the observed test statistic with the p-value (*Null Hypothesis Significance Testing*)

Sample Estimate and 95% CI

Theory:

If $y \sim Bin(n, \theta)$ in the population and y_i (0 or 1) are random and independent observations from the population, then the parameter θ can be estimated by maximising the likelihood function $L(\theta|y)$.

Once the data y are observed (i.e., y is known or fixed), the joint probability of the data y as a function of θ is called the likelihood $L(\theta|y)$. The $L(\theta|y)$ represents the probability that the outcome y is observed when the true value of the parameter is θ. In other words, $L(\theta|y)$ is a function of θ and it is equal to the probability distribution over fixed y for a given value of θ.

Using the maximum likelihood (ML) estimation method, we can maximise the likelihood function. The method identifies the highest joint probability of the observed data in terms of the parameter θ and represented by $\hat{\theta}$. For the binomial distribution, the ML estimate is given by: $\hat{\theta} = \sum_{i=1}^{n} y_i / n$.

The ML method estimates the parameter for which the observed data are most probable.

Figure 2c presents the graphical presentation of the likelihood function of the parameter given the observed data and corresponding 95% CI.

Null Hypothesis Significance Testing (NHST)

Theory:

If $y \sim Bin(n, \theta)$ in the population and y_i (0 or 1) are random and independent observations from the population, we can formulate a testable statistical hypothesis based on the observed data.

To test the null hypothesis (H_0: $\theta = \theta_0$) against the alternative hypothesis (H_1: $\theta \neq \theta_0$), we can compute a test statistic that summarises the data: $z = (\theta - \theta_0) / \sqrt{\theta_0 (1 - \theta_0) / n}$. *For large n, z is approximately normally distributed.*

We calculate the probability of obtaining the test statistic at least this extreme if the null hypothesis ($H_0 : \theta = \theta_0$) is true. If the probability (p-value) is smaller than a threshold (type 1 error, α), we reject the H_0.

Figure 2d presents the graphical presentation of the distribution of *test statistic*, the value of the *observed test statistic*, the *critical value* for the given α (commonly <0.05), and the *p-value* (*two-tailed*). It is important to emphasise that the *p-value* does not provide the direction, magnitude, and precision of the estimated effect.

5.3 Activities

- Change the sample size and discuss the effect of sample size on the mean and standard error of the estimate, the precision of confidence interval, test statistic, *p*-value.
- Assign different null hypotheses by considering different values of θ_0. Explain the outcomes of the hypothesis testing scenarios. Explore this further by changing the sample size.
- Change the significance level of the test. Explain the outcomes of different hypothesis testing scenarios.

6 Bayesian: *Simulate*

Figure 3 presents a screenshot of input (A) and outputs (B–D) for the *Bayes Simulate* section. In the simulation-based approach, we explore the theoretical basis of Bayesian methodologies using the same problem statement discussed in the *Frequentist* section.

6.1 Inputs

Parameter
Set a value of the *parameter* (θ) of the *binomial distribution* ($0 \leq \theta \leq 1$) to simulate the data.

Sample Size (*n*)
Select the *sample size* (*n*).

Prior: Parameters of Beta Distribution
Select the *shape parameters* of the *beta distribution* (α, β) (also called *hyperparameters*). See the next section for further explanation of the beta distribution as a prior distribution.

Click **Simulate** to simulate *Bayesian* analysis.

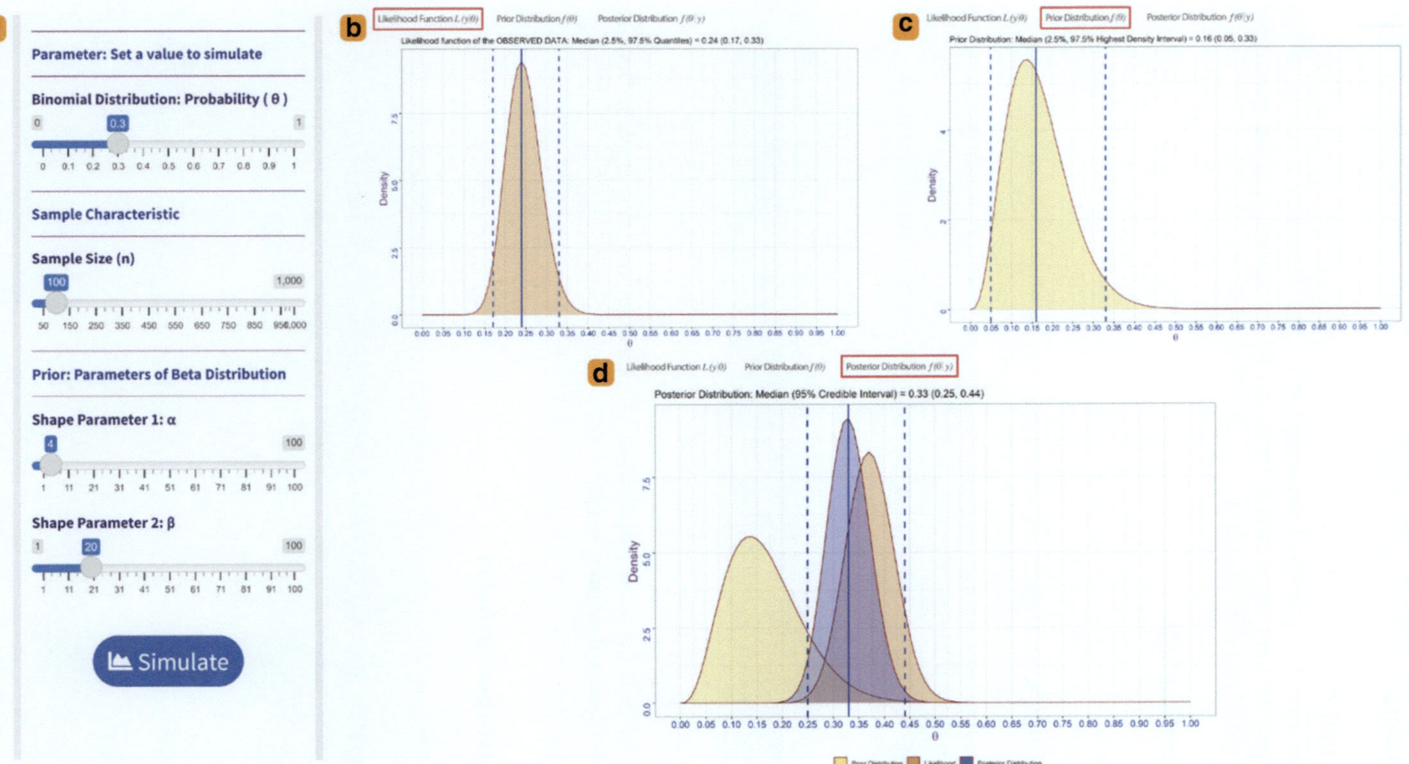

Fig. 3 Screenshot of *Bayesian Simulate* section showing (**a**) the input panel to set the simulation parameters of the data and prior distribution; (**b**) the likelihood function of the parameter given the observed data (*Likelihood Function*), (**c**) the beta prior distribution using the input values of shape parameters (*Prior Distribution*); (**d**) the beta posterior distribution employing the Bayesian theory (*Posterior Distribution*)

6.2   Outputs

Likelihood $L(\theta|y)$

Figure 3b presents the graphical presentation of the *likelihood function* of the parameter given the data and median, 2.5% and 97.5% quantiles of the parameter based on the simulated data of size n.

Prior Distribution $f(\theta)$

Theory:

Before observing the data, a fundamental component of Bayesian inference requires the prior distribution about the parameter θ.

The parameter θ in a Bayesian framework is not a fixed quantity; it follows a probability distribution. We provide our subjective belief about the parameter θ in the form of a probability distribution called prior distribution.

When the outcome variable follows a binomial distribution, a beta distribution—a continuous distribution represented by two shape parameters (α, β)—often captures the prior distribution. The beta prior is a conjugate prior since both prior and posterior distributions have the same distribution family (i.e. beta distribution).

Figure 3c presents the *prior distribution $f(\theta)$*, median, and 95% *Highest Density Interval (HDI)*. See explanations below.

Posterior Distribution $f(\theta|y)$

Theory:

If the probability distribution of the observed outcome y is $f(y|\theta)$ and the probability distribution of our prior belief about θ is $f(\theta)$, then according to Bayes rule, the posterior distribution $f(\theta|y)$ is

$$f(\theta|y) = f(y|\theta) f(\theta) / f(y)$$

The marginal probability of the data, $f(y)$, given the parameter space, does not depend on θ as the function integrates over all possible values of θ. We can drop the denominator $f(y)$ and represent the posterior distribution as:

$$f(\theta|y) \propto f(y|\theta) f(\theta)$$

As noted earlier, once the data y is observed, $f(y|\theta)$ is the probability of the data as a function of θ. It is called the likelihood $L(\theta|y)$. Hence, the posterior distribution is proportional to the likelihood times the prior distribution. In simple expression:

$$Posterior \propto Likelihood \times Prior$$

If the probability distribution of the observed outcome y is Bin(y|n, θ) and the probability distribution of the prior belief about θ is Beta(α, β), then the posterior distribution is given by a beta distribution with the following parameterisation:

$$Beta\left(\alpha - 1 + y, \beta - 1 + n - y\right)$$

The closed-form solution, as above, is not possible for most Bayesian problems, which involve multi-dimensional integrals of the marginal probability f(y). We use a simulation algorithm like Markov chain Monte Carlo (MCMC) to approximate the posterior distribution without computing f(y) (see Bayes Apply in the next section).

The 95% highest density interval (HDI) (also called credible interval) contains the parameter value with 95% probability.

The 95% HDI of the posterior distribution is generally narrower compared with the 95% HDI of the prior distribution; this reflects reduced uncertainty and increased precision of estimation with the observed data.

Note the fundamental conceptual difference and interpretation of 95% CI in the frequentist framework and 95% HDI in the Bayesian framework. In the frequentist view, the parameter is fixed (cannot have a distribution) and the CI is random (as it is calculated from the random sample). In the Bayesian view, the parameter has its own distribution and HDI is a concept in the context of the prior distribution and data.

Figure 3d presents the *posterior distribution f(θ|y)* median and *95% HDI*.

6.3   Activities

- Change the sample size and discuss the effect of sample size on the median and 95% HDI in the context of the likelihood and posterior distribution.
- Assign different values of hyperparameters of the prior distribution: ($\alpha = \beta = 20$); (α *low*, β *high*); (α *high*, β *low*). Explore the effect of the prior distribution on the posterior distribution.
- Explore the above scenarios with different sample sizes.
- A flat beta distribution prior can be defined with hyperparameters ($\alpha = 1, \beta = 1$). Explore the effect of a flat prior on the posterior distribution.

7 Bayesian: *Apply*

Figure 4 presents a screenshot of inputs (A) and outputs (B, C) for the *Bayes Apply* section. As mentioned earlier, the numerical solutions of the marginal probability for most Bayesian problems are intractable. Hence the Bayesian problem is commonly implemented using the MCMC method. We apply the Bayesian methodology to analyse an experimental dataset and estimate the population parameter with

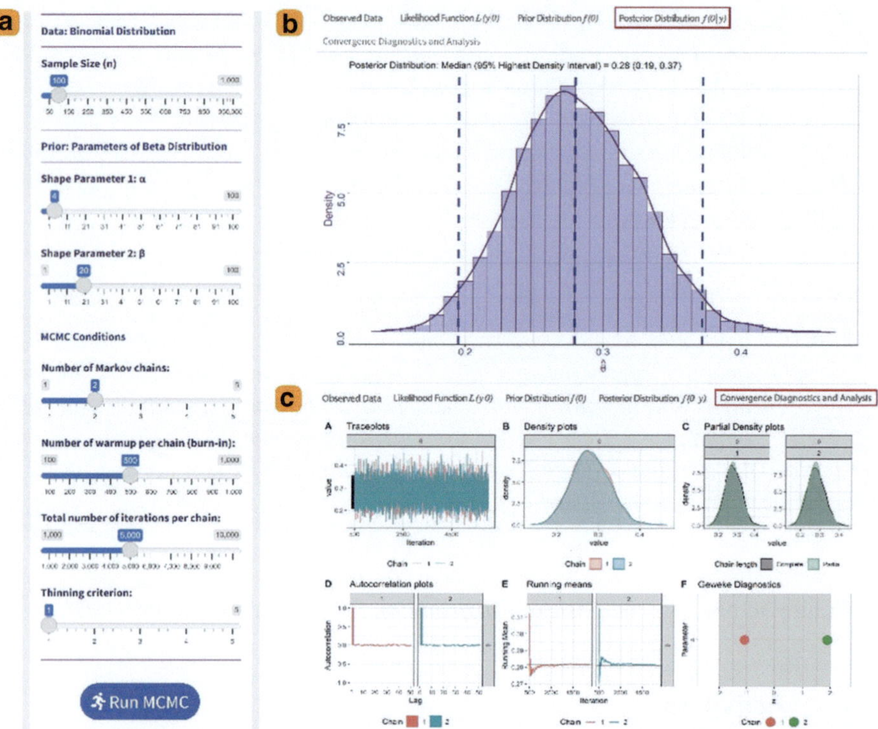

Fig. 4 Screenshot of *Bayesian apply* section showing (**a**) the input panel to select the sample size, parameters of the beta prior distribution and MCMC conditions; (**b**) the posterior distribution of the parameter using the MCMC method (*Posterior Distribution*); (**c**) summary checks of the Markov chain convergence to the target distribution using CODA tool (*Convergence Diagnostic and Output Analysis*). The figure does not present other tabs (*Observed Data, Likelihood Function, Prior Distribution*), similar to the *Bayes Simulate* section but specific to the input data provided in the input panel

associated uncertainty. There are several established software packages like WinBUGS [12], JAGS [13], and Stan [14] that implement complex and advanced Bayesian models. The application shows here a simple representation to understand the methodology and implement the MCMC tool to estimate the prevalence of the disease using the observed data.

7.1   Inputs

Sample Size (*n*)
Select the *sample size (n)*.

Prior: Parameters of Beta Distribution
Select the *shape* parameters (*hyperparameters*) of the *prior distribution*, i.e., *beta distribution*.

MCMC Conditions
To implement MCMC, we set some conditions: number of Markov chains, number of draws to discard as warmup (burn-in) per chain, number of iterations per chain, and thinning rate.

Click **Run MCMC** to apply the Bayesian model using the MCMC on the observed dataset.

7.2   Outputs

Observed Data
It displays the observed and summary data similar to the *Frequentist Apply* view.

Likelihood Function $L(\theta|y)$
It shows the *likelihood function* and median, 2.5% and 97.5% quantiles of the parameter based on the observed data of size *n*. It is similar to the *Bayes Simulate* section but specific to the input data provided in the Input panel.

Prior Distribution $f(\theta)$
It presents the *prior distribution* (*beta*) and 95% HDI based on the *shape* parameters chosen for the *beta distribution*.

Posterior Distribution $f(\theta|y)$

Theory:

One approach to approximate the posterior distribution in a Bayesian framework is to use the MCMC method [15]. It is implemented by constructing a Markov chain, formed by a sequence of random events, in which the probability of each event depends only on the state attained in the previous event. Markov chains create an equilibrium distribution, and therefore, MCMC methods are used for calculating the numerical approximation of posterior distribution. This method is the most reliable since most complex Bayesian problems involve multi-dimensional integral where closed-form solutions do not exist [16].

The MCMC method used two common algorithms: Metropolis–Hastings (MH) algorithm [17] and the Gibbs sampler [18]. The latter one is a special case of MH and is implemented here.

Often, the precise form of the posterior distribution is unknown. The MCMC method, using the Gibbs sampling or MH algorithm, generates a large number of samples from the posterior distribution to cover the entire distribution.

The application employs Gibbs sampling, which generates a multi-dimensional Markov chain of the random variable θ [16].

The simple scenario here (only one parameter) starts with an arbitrary starting value of θ, draw a new value of θ from the conditional distribution given the starting value and the data. We then repeat the cycle many times.

In general, for a vector of unknown θ, say $\theta = [\theta_1, \theta_2]$, it chooses arbitrary starting values for each component $\theta_1^{(0)}, \theta_2^{(0)}$. It then samples a new $\theta_1^{(1)}$ from the full conditional distribution θ_1 given the most recent values of other elements of θ and the data y:

$$\theta_1^{(1)} \sim f\left(\theta_1|\theta_2^{(0)}, y\right)$$

Next, sample a new $\theta_2^{(1)}$ given $\theta_1^{(1)}$ and y as:

$$\theta_2^{(1)} \sim f\left(\theta_2|\theta_1^{(1)}, y\right).$$

This completes one iteration of the Gibbs sampler and repeat the above step many times (T).

We generally drop the first b samples as a warm-up to stabilise the chain, referred to as 'burn-in'.

Each cycle of T samples is called a 'chain', and it is often advised to run multiple chains.

We obtain the summary statistics (median and quantiles) from all realisations from the marginal posterior distributions of θ.

Figure 4b demonstrates the graphical representation of the *posterior distribution* $f(\theta|y)$ from the MCMC sampling, median, and 95% HDI.

Convergence Diagnostics and Output Analysis (CODA)

Theory:
In the MCMC method, it is essential to check that the Markov chain converges to the target distribution, i.e., the posterior distribution. We check the convergence of the Markov chain using the convergence diagnostics and output analysis (CODA) tool.
For example, the trace plot captures all the realisation of MCMC samples with the x-axis displaying the iteration and the y-axis the parameter's value.

Figure 4c presents a selected set of graphical displays of convergence diagnostics. These include trace plots, density plots, partial density plots, autocorrelation plots, running means, and Gweke's diagnostics for all the chains. Detailed interpretations of these plots and other CODA statistics are available in the literature [19, 20]. Some formal convergence diagnostic statistics and plots commonly used are Gelman and Rubin, Gweke, Raftery and Lewis, Heidelberger and Welch [21].

The current version of the application does not present the Bayes factor, the Bayesian alternative to NHST. The *Bayes factor* of two competing hypotheses (null versus alternative) is defined as the ratio of the posterior odds to the prior odds. The concept of *posterior predictive distribution*, which is the distribution of possible unobserved values conditional on the observed values, is not covered either. The extension of the Bayesian application may also include the scenario of disease prevalence and the sensitivity and specificity of a diagnostic test. A future version of the application will address these additional concepts.

7.3   Activities

- Explore the activities presented in the *Bayes Simulate* section.
- Explain different MCMC conditions; change the conditions and explore how the posterior distribution and the convergence diagnostics change.

8   Lesson Plan

The lesson plan and teaching strategies can introduce a single framework or integrate frequentist and Bayesian frameworks with appropriate objectives, discussions, and activities. The target students may include beginner, intermediate, and advanced levels. At the beginner level, the goals are to comprehend the concept, describe the principles and approach to solve a problem, and interpret the outcomes. The intermediate students should examine the problem scenario, identify and illustrate the solution with statistical formulations, and evaluate the results with background statistical principles. At the advanced level, students should construct the solutions using simulation. They also should learn beyond the application, implement and explore other software tools to resolve this and more complex problems. Advanced learners can review the source code of the application for additional help. We present a list of intended learning objectives and topics that can be integrated with the application. Topics at an advanced level and those not covered fully in the current application are indicated with an asterisk (*).

8.1 Frequentist Framework

Intended Learning Objectives
- Explain the concept of frequentist inference (Beginner)
- Describe fundamentals of inferring population parameter using frequentist inference: confidence interval and null hypothesis significance testing (Beginner)
- Identify a one-parameter problem to infer the population parameter, illustrate all steps (Beginner, Intermediate), formulate the mathematical framework (Intermediate), generate sample data, and perform the analysis (Beginner, Intermediate, Advanced)

Topics
Frequentist inferences: sample, population, estimate, parameter, relative frequencies, probability, probability distributions, binomial, normal and standard normal distributions, sampling distribution, random sampling, sample mean, standard error of mean, unbiased estimate*, central limit theorem*, confidence interval, likelihood function*, maximum likelihood method*, Fisher information*, null hypothesis significance testing (NHST), steps of NHST, statistical hypothesis, null and alternative hypothesis, type 1 error, type 2 error*, significance level, critical region, test statistic, z-statistic, p-value, power of a test*.

8.2 Bayesian Framework

Intended Learning Objectives
- Explain the concept of Bayesian inference (Beginner)
- Describe fundamentals of inferring the population parameter using Bayesian inference: likelihood function, prior and posterior distribution (Beginner)
- Discuss the approach to the application of Bayesian inferences using MCMC methods (Beginner)
- Summarise the similarities and differences between frequentist and Bayesian inferential frameworks (Beginner)

- Identify a one-parameter problem to infer the population parameter, illustrate all steps (Beginner, Intermediate), formulate the mathematical framework (Intermediate), generate sample data, and perform the analysis using own scripts or in a standard software environment (Beginner, Intermediate, Advanced)

Topics

Bayesian inferences: probability, conditional probability, marginal probability, discrete and continuous distributions, probability density function, Bayesian theorem, likelihood, prior distribution, conjugate families of priors*, hyperparameters, posterior distribution, posterior predictive distribution*, highest density interval (credible interval), region of practical equivalence (ROPE)*, Bayes factor*, MCMC conditions, Metropolis–Hastings algorithm*, Gibbs sampler*, convergence diagnostics and analysis*.

9 Conclusions

The paper presents a simple, user-friendly, interactive application with underlying theories to demonstrate key concepts of frequentist and Bayesian statistics with an example of estimating the prevalence of disease in a population. It considers the distribution of the disease state as a binomial distribution and the subjective knowledge of the prevalence as a beta distribution (a conjugate prior). The article illustrates methodologies, outcomes, and interpretations from both inferential methods intertwined with the application interface. Educators can integrate the application with specific learning objectives, lesson plans, and activities for beginner, intermediate, and advanced learners.

Acknowledgement The author thanks two anonymous reviewers and editors for their constructive suggestions. The author also thanks Wai-Lum Sung to create the figures.

References

1. Hoegh A. Why Bayesian ideas should be introduced in the statistics curricula and how to do so. J Stat Educ. 2020;28(3):222–8. https://doi.org/10.1080/10691898.2020.1841591.
2. Wasserstein RL, Lazar NA. The ASA's statement on p-values: context, process, and purpose. Vol. 70, American Statistician. Alexandria: American Statistical Association; 2016. p. 129–33. https://doi.org/10.1080/00031305.2016.1154108.
3. Wasserstein RL, Schirm AL, Lazar NA. Moving to a world beyond "p < 0.05." Vol. 73, American Statistician. Alexandria: American Statistical Association; 2019. p. 1–19. https://doi.org/10.1080/00031305.2019.1583913.
4. Hu J. A Bayesian statistics course for undergraduates: Bayesian thinking, computing, and research. J Stat Educ. 2020;28(3):229–35. https://doi.org/10.1080/10691898.2020.1817815.
5. van de Schoot R, Kaplan D, Denissen J, Asendorpf JB, Neyer FJ, van Aken MAG. A gentle introduction to Bayesian analysis: applications to developmental research. Child Dev. 2014;85(3):842–60. https://doi.org/10.1111/cdev.12169.

6. Kruschke JK, Liddell TM. The Bayesian new statistics: hypothesis testing, estimation, meta-analysis, and power analysis from a Bayesian perspective. Psychon Bull Rev. 2018;25(1):178–206. https://doi.org/10.3758/s13423-016-1221-4.

7. Chang W, Cheng J, Allaire J, Sievert C, Schloerke B, Xie Y, et al. Shiny: web application framework for R. 2021. Available from https://cran.r-project.org/web/packages/shiny/.

8. Nath M. ABACUS: Apps based activities for communicating and understanding statistics. 2019. Available from https://cran.r-project.org/web/packages/ABACUS/index.html.

9. R Core Team. R: A language and environment for statistical computing. R Foundation for Statistical Computing, Vienna, Austria. 2021. Available from https://www.r-project.org/.

10. Albert J, Hu J. Bayesian computing in the undergraduate statistics curriculum. J Stat Educ. 2020;28(3):236–47. https://doi.org/10.1080/10691898.2020.1847008.

11. Bland M. An introduction to medical statistics. 4th ed. Oxford: Oxford University Press; 2015.

12. Lunn DJ, Thomas A, Best N, Spiegelhalter D. WinBUGS - A Bayesian modelling framework: concepts, structure, and extensibility. Stat Comput. 2000;10(4):325–37. https://doi.org/10.1023/A:1008929526011.

13. Plummer M. JAGS - just another Gibbs sampler. 2017. Available from https://mcmc-jags.sourceforge.io/.

14. Carpenter B, Gelman A, Hoffman MD, Lee D, Goodrich B, Betancourt M, et al. Stan: a probabilistic programming language. J Stat Softw. 2017;76(1):1–32. https://doi.org/10.18637/JSS.V076.I01.

15. Gelfand AE, Smith AFM. Sampling-based approaches to calculating marginal densities. J Am Stat Assoc. 1990;85(410):398–409. https://doi.org/10.1080/01621459.1990.10476213.

16. Lunn D, Jackson C, Best N, Thomas A, Spiegelhalter D. The BUGS book: a practical introduction to Bayesian analysis. In: The BUGS book. London: Chapman and Hall; 2012. https://doi.org/10.1201/B13613.

17. Chib S, Greenberg E. Understanding the Metropolis-Hastings algorithm. Am Stat. 1995;49(4):327–35. https://doi.org/10.1080/00031305.1995.10476177.

18. Casella G, George EI. Explaining the Gibbs sampler. Am Stat. 1992;46(3):167–74. https://doi.org/10.1080/00031305.1992.10475878.

19. Cowles MK, Carlin BP. Markov Chain Monte Carlo convergence diagnostics: a comparative review. J Am Stat Assoc. 1996;91:883–904.

20. Roy V. Convergence diagnostics for Markov chain Monte Carlo. Annu Rev Stat Appl. 2020;7:387–412. https://doi.org/10.1146/annurev-statistics-031219.

21. Plummer M, Best N, Cowles K, Vines K, Sarkar D, Bates D, et al. Coda: output analysis and diagnostics for MCMC. 2020. Available from https://cran.r-project.org/web/packages/coda/index.html.

Teaching Data Analysis to Life Scientists Using "R" Statistical Software: Challenges, Opportunities, and Effective Methods

Renata Medeiros Mirra, Jim O. Vafidis, Jeremy A. Smith, and Robert J. Thomas

1 Introduction

1.1 Why Do (Life-) Scientists Need Statistics?

A good understanding of statistics enables a researcher to address a vast array of scientific question in a rigorous and quantitative manner, by collecting the right sort of data in well-designed experiments, by using the most appropriate statistical tests, and by interpreting the results correctly [1]. The central importance of these tasks to the scientific process means that the ability to understand, use and interpret statistics is one of the most fundamental and empowering skills that a scientist can possess.

Transferable skills are arguably the most important set of skills that students will gain in higher education. Statistics is an essential tool for quantitative thinking, and as a central discipline in many academic and non-academic careers is therefore an important transferable skill to be developed by life science students [2]. For example, in the UK higher education system, statistical skills occupy a prominent place in the Biosciences Quality Assurance Accreditation Benchmarks [3], both as a core biosciences skill and as a key transferable skill, and is an important component of

R. Medeiros Mirra (✉)
Eco-explore Community Interest Company, Cardiff, UK

Cardiff School of Dentistry, Cardiff University, Cardiff, UK
e-mail: medeirosmirrarj@cardiff.ac.uk; https://www.eco-explore.co.uk

J. O. Vafidis
Eco-explore Community Interest Company, Cardiff, UK

University of the West of England, Bristol, UK

J. A. Smith · R. J. Thomas
Eco-explore Community Interest Company, Cardiff, UK

Cardiff School of Biosciences, Cardiff University, Cardiff, UK the knowledge and skills that new medical and dental graduates should demonstrate [4, 5]. Nevertheless, there has been—and still is—substantial variation between and within universities in the UK, in the type and level of statistical training available for undergraduates in the life sciences [6, 7].

1.2　Teaching and Learning Statistics

Despite the central importance of data analysis to the day-to-day nature of research, we observe that many life scientists, including many academic university staff as well as postgraduates and undergraduates, feel ill-equipped and/or exhibit a reluctance to engage with statistical analysis, as either learners, practitioners, or teachers [8]. As a result of this reluctance, statistical training is often "out-sourced" by academic staff, e.g. by asking others to provide support for data analysis in final-year undergraduate projects. Where it occurs, this lack of engagement with statistics by academic staff is detrimental because a lack of statistical skills compromises a researcher's independence, limits the types of questions that they are able to address, limits their understanding and critical evaluation of others' work, and seriously constrains the range of teaching and research support that they can provide to students [9].

Evidently, statistical analysis does not necessarily come easily to many life scientists [8], but the research world has developed over recent decades such that data analysis has become an increasingly important and useful part of the toolkit of techniques that are available for understanding the living world. Indeed, it could be argued for many research fields that without data analysis a modern scientist is primarily a data collector, or a research manager and a (sometimes poor) interpreter of other people's research. This makes them completely reliant on collaboration with data analysts, who are usually detached from the development of the research questions and are often consulted late in the process after data has already been collected. We strongly believe that the scientific process would be improved by a greater awareness of statistical issues among the wider research community.

In contrast to many academic staff, most current life science students are expected (and usually compelled) to engage with statistical analysis in a wide range of contexts throughout their training, at undergraduate and postgraduate levels. Many students find this engagement with statistics highly challenging; for example, difficulty understanding the meaning and interpretation of p-values has been reported as a major problem among biosciences students [10], who often fail to recognise the relevance of statistics to their wider studies. This is the experience of many students across the life sciences, although those who manage to become competent and confident in data analysis gain important transferrable skills which they will employ for the rest of their scientific careers, and which greatly boosts their employability. In the words of a first-year biosciences undergraduate at Cardiff University: "Doing stats makes me feel like a proper scientist".

This current mismatch between the level of statistical training and support for data analysis offered by many academics in the life sciences, and the great need for such training at undergraduate and postgraduate level, poses a major challenge as academic staff feel unable or unwilling to provide high-quality statistical teaching and/or support to their students. One likely reason for this mismatch having arisen is the rapid development of statistical methods and software over recent decades. The speed at which new statistical methods now become accessible to practitioners, such as life scientists, makes it difficult for busy lecturers with many different responsibilities to keep up with current methods.

There is a wide range of different software used in statistical teaching, with many students being taught in SPSS [11], but increasingly R has been the preferred software for researchers and for teaching statistics in the life sciences, particularly for postgraduates [12–14]. Here we reflect on the advantages, challenges, and opportunities of using R in university-level teaching and learning to both undergraduates and postgraduates. We draw on our experiences (and extensive student feedback) collected at Cardiff University School of Biosciences across three academic years between 2014 and 2017. We draw also on experiences of teaching independent courses run by the Eco-explore Community Interest Company (www.eco-explore. co.uk) since 2014, which is aimed at PhD researchers in the UK. Cardiff School of Biosciences enrols over 300 undergraduates every year, who graduate into one of eight possible degree courses: biochemistry, genetics, neurosciences, anatomy, physiology, biomedical sciences, biological sciences, and zoology. Most students will have not taken Mathematics for A-level and their backgrounds and career aspirations are diverse. Examples include careers in academia, in industry or in the third sector, training as teachers, engaging in medical training or in careers in the medical sector.

1.3   What Is R?

R is a very powerful and flexible statistical software package that is suitable to clean, analyse, visualise, and present more or less any type of quantitative data and is compatible with all of the common computer operating systems (Windows, Mac, and Linux). R is actually a computing environment and programming language, rather than a statistics package in the usual sense; unlike many of the familiar statistics packages commonly used in undergraduate teaching (e.g. Minitab, SPSS), the user tells R what to do primarily by typing in commands, rather than clicking on options in a menu (Fig. 1).

R is a relatively new phenomenon, having been developed from the S and S++ statistical programming language(s), and since the launch of R in the 1990s [15] it has been widely adopted across many scientific disciplines and commercial contexts [16]. Importantly, and remarkably, R is completely free to download and use (https://cran.r-project.org/) and is well supported by an enthusiastic online community of statisticians, applied researchers, and education practitioners.

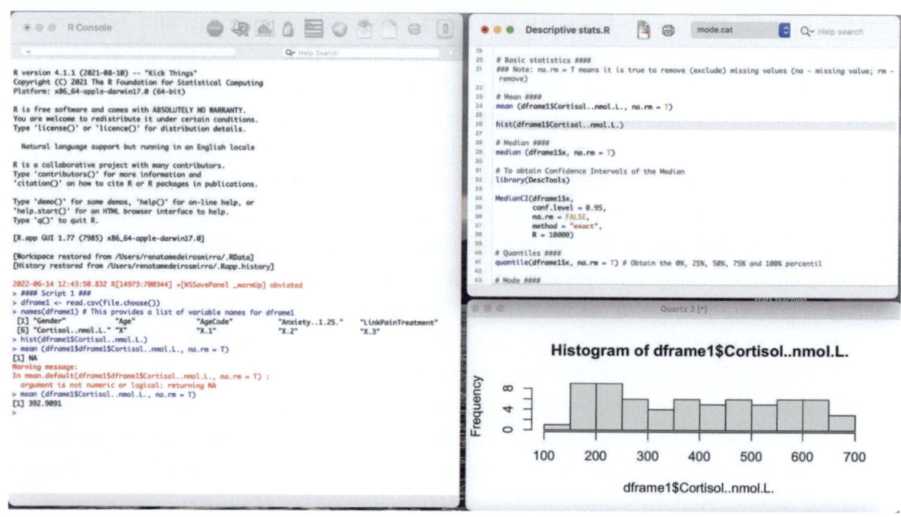

Fig. 1 Interface of R statistical software basic installation

Table 1 Online discussion forums and blogs for obtaining ideas and practical help with R

Forum name	Web address	Audience
R-Space	Search for "R-Space" in Facebook	R-learners at any level, simple questions welcome
R-bloggers	https://www.r-bloggers.com	R news and tutorials contributed by hundreds of R-bloggers
Cross-validation	https://stats.stackexchange.com	A statistical community within the "Stack Exchange" forum
Stack overflow	https://stackoverflow.com	A question and answer site for professional and enthusiast programmers

Many of the world's applied statisticians now use R. Because R is "open source", many statisticians contribute new data analysis methods in the form of "packages" that are added to the "base installation" of R and anyone can download and use these packages for free. Many of these packages implement methods from recent statistical research. Consequently, R is generally more up-to-date than traditional statistical programmes as there is less lag-time in development and its functionality can be expanded dramatically as soon as each new method becomes available. Such accessibility to new statistical developments itself has potential disadvantages because users may put unwarranted faith in packages that may implement provisional or speculative methods and so which may still be under development [14, 17]. Fortunately, the online community of R-users provides an effective form of peer review and quality control. New methods are tested and critiqued by them via online forums (e.g. those listed in Table 1), as well as in the primary research literature [12].

Analyses are normally implemented in R from a "script", which is essentially a programme for carrying out a set of commands in R. The script can be saved as a file, which can be used to re-run the analysis, shared with others, and edited if desired to modify the analysis. The script can also be used to carry out the same actions on other data, simply by altering the names of the data file or variables. As activities undertaken in R start to become more complicated, this becomes very useful, minimising typing errors, and saving time. In fact, once learners have built up a set of annotated script files, most analyses can be carried out by copying, pasting, and editing pre-existing scripts.

Partly because of the advantages outlined above, and no doubt partly because R is free, many universities are attracted to R in preference to commercially available software that incurs substantial licensing costs. R is becoming increasingly widely adopted across many life science and social science disciplines, including in the biosciences, where it is now one of the most widely cited data analysis software programmes [12, 13, 16, 18].

Importantly, R provides a means of archiving and sharing scripts, allowing open and direct access to the detail of statistical analyses presented in research publications, as outlined below. Despite these important advantages of R (some of which are shared with other software that offer command-line options, e.g. SAS, STATA, SPSS), users more comfortable with menu-driven software packages often feel reluctant to make the transition into using R because of its seemingly challenging command-line format and programming language.

1.4   Climbing the R Learning Curve

R has a reputation of being challenging to learn, though this seems to be a greater challenge for users who are already familiar with menu-driven statistical software. Perhaps surprisingly, our experience is that undergraduates learning about data analysis for the first time do not necessarily find it any more difficult to learn statistics using R than using any other software [11]. Indeed, the main difficulties identified by students when using R ("confusing", "hard to learn", and "not user-friendly") are common to other software packages, and to learning statistics in general. In contrast, postgraduates who were previously taught using a menu-driven software package seem to find R more challenging at first. Nevertheless, the process of using the command-line format of R provides important programming skills that all students appreciate are transferable skills that they can apply in contexts beyond statistical hypothesis testing (e.g. producing publication-quality graphics).

Furthermore, it can be argued that command-line programming of analyses promotes a deeper comprehension about the statistical methods themselves, essentially because a greater level of understanding is generally required to successfully write or edit a line of code than to click on a menu option. This is not a given however, and as pointed out by Shuker [14], one of the great challenges in teaching R is to ensure that students understand the meaning of the code that they are using, rather than using scripts as "recipes", without engaging fully with understanding what the commands mean. One way to avoid this is to encourage students to annotate their scripts in detail, e.g. by using hashtag notes [1]. Nevertheless, some feedback comments by students learning statistics through R do suggest that they feel they are blindly following the script, e.g. "there is no point being given code, and told what it does (when it works) if we are not told how to manipulate it. Lectures are not appropriate only computer-based workshops would be appropriate" or "Script huge help. Don't really know how to use R, just followed script. Won't remember it off by heart. No idea how to carry out functions". By contrast, other students seem more able to benefit from deeper understanding that the command software can offer, e.g. "More fun as I progressed and understood it. R make stats more accessible". or "stats is really easy if we know how to read R, which is the challenging part and exciting".

The process of understanding, writing, and annotating code makes explicit not only the choice of overall analysis method, but also the choice of specific aspects of the method, as defined by the model structure and the "arguments" to the statistical function being applied (e.g. choice of independent variables, interactions, structure of the error term, error family, link function, dispersion parameter, offset and weighting variables, missing values, etc.). In our experience, as students learn how to visualise the patterns in their data, design their own analyses, and implement these analyses using annotated R-scripts, they come to understand their datasets, and to appreciate the close connection between their research questions and model design.

The net result is that some students are excited to learn R, even if they are not very excited to learn about statistics, because in doing so they are becoming programmers as well as data analysts. This is illustrated in feedback comments like: "I like playing with R, less so the stats! "Any chance of using Python?" or "Like R, not a big fan of statistics". Student comments about their experience of learning statistics through R also frequently reflect their awareness of the relevance and value of the programming skills that they are simultaneously learning. Other students express a dislike for the topic in general and it is unlikely that using a different software would improve this. Overall, we believe that the benefits of learning R, and using it as a powerful tool for data analysis, generally offset the relatively steep learning curve experienced by at least some learners as they get started with using the software. However, it is true that some students particularly dislike the programming element and find learning R stressful. In those cases, using R may add to the challenge of learning statistics, rather than enhancing the learning process [17], even among those who like statistics to start with.

In an ideal scenario, different students could be given different software options to try and decide on the one best suited to their needs and aspirations. However, this is largely unpractical from a teaching point of view, considering the many limitations in terms of time and resources that most statistics teachers face, often teaching very large cohorts of students with limited timeframes. Hence, the choice of statistical software (if any) for teaching statistics must weigh the benefits and demerits across the whole cohort. Specifically, when considering teaching statistics using R, we focused on whether (1) it enhances the learning of statistics to a greater number of students compared to those to whom it causes added difficulties; (2) the benefits are themselves more significant than the disadvantages (e.g. whether the scale of increase in engagement and enjoyment surpasses the scale of increase in stress); and (3) if the longer-term benefits surpass the shorter-term challenges. In fact, a very common theme in feedback comments from students is that R is difficult at the start and can take time, but makes things easier in the longer term.

In this context, it needs to be acknowledged that in settings where the relevance of statistics is less obvious to the students, for example, in medical contexts, and/or where statistics are taught over very short timeframes, overcoming the R steep learning curve might be more challenging or may simply not be possible. In such contexts, engagement with statistics may be improved by using menu-driven "front-end" software for using R, which can prime students to deeper engagement with the command-line aspects of R in the future. Examples of such "front-end" software, which are also free, include JASP [19] (Fig. 2) and jamovi [20] (Fig. 3). These two software packages have in common with R the philosophy that scientific software should be "community driven", allowing users to contribute to its development and improvement, and encouraging communities of users to support each other. Furthermore, all the analyses available in jamovi are available from within R using the Selker et al. [21], and jamovi can be placed in "syntax mode", producing the R syntax required to reproduce jamovi analyses in R. Alternatively, the user can enter

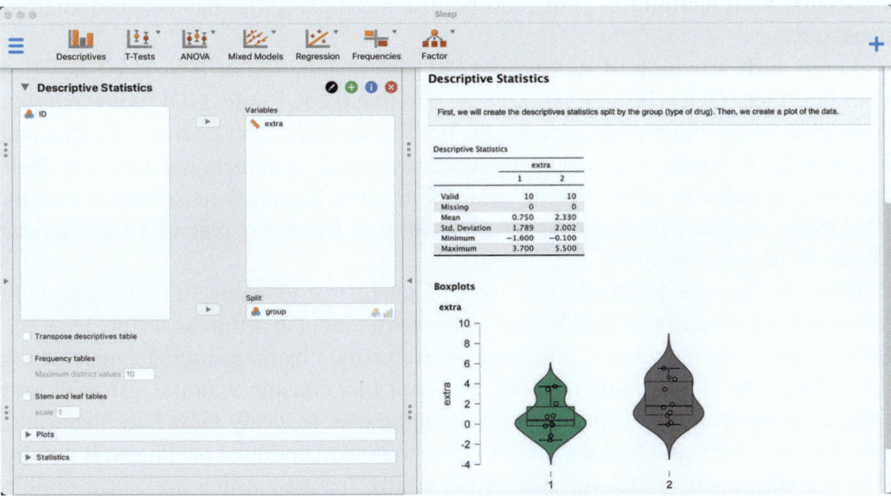

Fig. 2 Interface of statistical software JASP

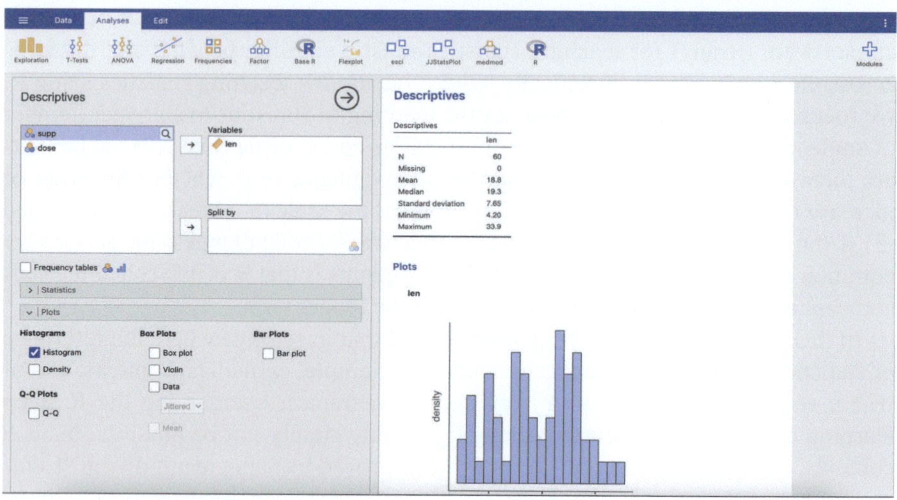

Fig. 3 Interface of statistical software jamovi

R-code, and analyse data using R within jamovi. Similarly to what SPSS or STATA provide, JASP has focused on producing an interface that retains the desirable properties of syntax (such as reproducibility) without requiring or producing syntax itself. The JASP team are now also in the process of adding the syntax functionality to the software, that enables the user to create a direct link to R. While jamovi or JASP can offer a good trade-off between functionality and accessibility, some of the advantages of R previously discussed, including acquiring the transferable skills associated with learning to code, can be lost when using the menu-based alternatives to R.

In our early teaching of R, we used the base installation of R as our Graphical User Interface (GUI) (Fig. 1), but we now find that the R-Studio GUI (www.rstudio.com) offers some important advantages, both for teachers and learners. Firstly, and importantly, R-Studio gives all users the same format on screen, regardless of their operating system. In other words, what the teacher has on their screen is exactly what every student will see, and this aids with step-by-step explanations during our lecture-workshop sessions.

Other features of R-Studio that we find useful for teaching include automatic colour-coding of script, which makes it easy to distinguish different script elements, such as functions, arguments, notes, etc. R-Studio has clearly arranged panels which make it easy to view simultaneously the script file, console window, graphics and computing environment (Fig. 4). Tabs allow the user to easily view help materials, data-frames, command history, and to view, download and load packages. R-Studio also has very useful indexing and bookmarking features which are not available with R's base installation GUI. R-studio is now so widely used that if feels the more useful platform to introduce students to.

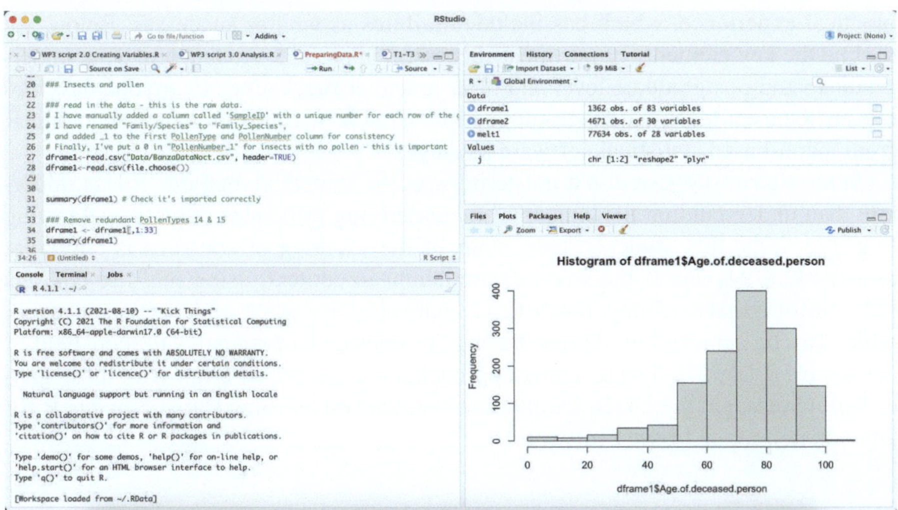

Fig. 4 Interface of R-studio for R statistical software

2 Teaching Statistics Using R

The introduction of statistical computing to university life sciences departments transformed the teaching of data analysis from the early 1990s onwards. At first, software-based statistics courses, both at undergraduate and postgraduate level, focused on implementing the "classical" statistical tests (e.g. t-tests, regressions, ANOVA and their non-parametric equivalents). From the turn of the millennium, such courses—particularly at postgraduate level—often included relatively more advanced methods such as General Linear Models (GLMs/ANCOVAs), Generalised Linear Models (GLMs with non-Gaussian error families), and Generalised Linear Mixed Models (GLMMs, combining fixed and random independent variables), as these methods became widely supported by the commercial software packages.

At the School of Biosciences at Cardiff University in Wales, UK, R has been taught at postgraduate level (Masters in Research in Biosciences and Doctor of Philosophy) to cohorts of ~15 to 40 students since 2006, and at undergraduate levels (1st year and 2nd year) to cohorts of ~350 to 420 students since 2012; thereby making Cardiff School of Biosciences a pioneer of teaching statistics through R, relative to most other UK university biological sciences departments. Prior to the transition to teaching R, data analysis was taught primarily using Minitab, though other packages (SPSS, Genstat, etc.) were also widely used for analysis at postgraduate level within individual research groups.

As well as the learning curve for R-learners, it is clear that there is also a steep learning curve for those tasked with teaching R to others. Our early attempts to teach R were, in hindsight, not always well-designed. Our teaching methods have evolved through a kind of trial-and-error-driven natural selection, in which we have retained more effective methods and discarded less effective ones. As a result, our overall approach and specific methods have changed substantially in the light of our practical experience, which has included failures as well as successes. Below, we outline the key elements of our most successful approaches to teaching R at undergraduate and postgraduate level and offer practical suggestions to anyone developing an R-based teaching programme from scratch or transitioning to teaching R from menu-based statistical software packages.

Studies have suggested that mastering specific statistical methods is less important than understanding the concepts and underlying principles of statistical testing (e.g. [22, 23]); that students learn better if the teaching of statistics is computer assisted [23, 24] and if the students are taught in context, using real-life datasets [25]; and also that teaching of statistics is more effective when students feel engaged, which can be achieved by demonstrating the relevance of statistics to their field of studies [6, 8, 26, 27]. The teaching approaches we developed align with these general principles, but there is little literature available on teaching statistics using R (or any other command-based software).

2.1   Methods for Teaching and Learning Statistics Using R

A key aspect of our teaching of R is the integration of statistical theory (e.g. sum of squares, central limit theorem, mathematical basis for statistical tests) with the practical application of the theory to example datasets, combined with substantial individual support for learners, particularly as they begin to implement analyses of their own datasets. This allows the theoretical teaching to be immediately applied to datasets that the learner is motivated to understand, and while the information is fresh in the learner's mind. This also promotes valuable discussion (e.g. "why is this method appropriate for this dataset?"; "why am I getting these error messages?"; "how do I interpret and report this 2-way interaction"?). The mix of statistical theory and practical application aims to give students both "a sense of the *structure* of the subject [rational aspect] and a sense of the *worthwhileness* of the subject [emotional aspect]" [28]. This "learning by doing" approach is implemented using a set of complementary methods, outlined below for undergraduates and postgraduates. Tables 2 and 3 outline the structure of an undergraduate (Table 2) and postgraduate course (Table 3) using R.

2.2   Undergraduate Teaching

Between 2012 and 2016, students at the Cardiff School of Biosciences were taught Data Analysis as part of a 20 credit Skills for Science module in year 1 via 12 lectures and nine optional drop-in Stats Clinics. The students were given one large dataset (real data collected on themselves through a survey during the first lecture) and had to complete three pieces of coursework using R to perform a range of statistical analyses on the data; initially all pieces of coursework were individual and summative assessments but, in response to students' feedback, the first coursework was made formative, and the second coursework was made of group work. The first piece of coursework included tasks that required cleaning the dataset (e.g. removing

Table 2 Structure of a typical undergraduate course in data analysis at Cardiff School of Biosciences using R statistical software. This material is typically taught in 12 weekly sessions, supported with clinics, online support, and coursework feedback

Core sessions (1 h each)	Topic
1	Introduction to research questions and statistics
2	Descriptive statistics
3	Normal distribution
4	Hypotheses testing
5	Choosing a statistical test
6	Categorical data—Chi-squared test
7	Testing for associations (correlation)
8	Testing for differences (t-test)
9	Revision Q&A
10	Introduction to linear regression
11	Introduction to ANOVA
12	Guest speaker speaking about "statistics in the real world"
Optional drop-in sessions to support coursework (2 h each × 3 per coursework)	
Coursework 1	Descriptive statistics, data distribution, and plotting using R
Coursework 2	Descriptive statistics, t-test, correlation, and chi-squared test using R
Coursework 3	Descriptive statistics, t-test, correlation, and chi-squared test, including paired and non-normal data, using R

Table 3 Structure of a typical postgraduate course in data analysis at Cardiff School of Biosciences that uses R statistical software. This material is typically taught in 2 × 2-hour core sessions per day, across a 5-day period, or 1–2 weekly sessions over a longer period

Core sessions (2 h each)	Topic
1	Getting started with R and RStudio; syntax, scripts, and data handling
2	The General Linear Model (GLM)—combining ANOVA and regression. Reporting your GLM visually and quantitatively
3	GLMs for count data
4	GLMs for binomial data
5	Avoiding pseudo-replication: generalised linear mixed models (GLMM)
6	Modelling wiggly lines with generalised additive models (GAMs and GAMMs)
7	Multi-model inference and model averaging
8	Multivariate methods: including principal component analysis (PCA), cluster analyses, discriminant function analysis
9	Time series analysis and survival analysis
10	Spatial analysis and mapping
Workshop sessions (1 h each)	
1	Investigating your own data (formatting and graphical exploration)
2	Investigating your own data (model building and graphing model predictions)
3	Messy datasets: models in the real world!
Optional sessions (0.5–1 h each)	
1	Dealing with small datasets
2	Circular statistics
3	Advanced graphics

impossible values or re-categorise variables). These tasks were kept relatively simple but arguably a greater focus could be put into teaching students more data cleaning and data handling, as these tasks can now be carried out very quickly with the development of R-packages like Tidyverse. Even though some students clearly did not put enough effort into the assessments (potentially because they didn't feel confident with the tasks) and received very low marks (below 10%), most students performed very well, with the mean and median scores typically greater than 75% and 80%, respectively. Each of the three pieces of coursework was supported by three drop-in Stats Clinics, where the students could carry out the set tasks, helped by the lecturer and a team of postgraduate demonstrators (teaching assistants). These clinics were key to the learning process and allowed the students to practice using R in a dedicated "safe space", where they could make mistakes and get immediate support from the teaching team and also from their peers. The clinics were mentioned extensively in feedback comments from students as helpful, or even essential, to the learning process and we are convinced they were a major contributor to enable students to consistently perform to such high standard.

In year 2, students were taught statistics as part of a module on Research Techniques. This included three statistics lectures at the start of the academic year: the first lecture was revision of topics from year 1, the second lecture covered the mathematical principles behind ANOVA and Linear Regression and the third lecture was on experimental design. Apart from these three formal taught sessions, the statistical component of the module was self-taught through resources posted online. These resources were primarily problems and answers to these problems, which allowed the students to test themselves and learn through problem-solving. At the end of the autumn semester, the students could perform a formative test for which they received detailed feedback. The students were assessed through an open book class test in the spring semester—the test consisted of a biological problem, with data to analyse using R, and questions to answer about the interpretation of the results. The students were told that the test was on ANOVA (Simple, Multiple, or Repeated Measures) and/or Linear Regression and they should bring their own annotated R-script with everything they need to complete the task. The mean and median exam marks across the three academic years reflected on this paper ranged between 60–65% and 66–70%, respectively. Each year, there were between 5 and 11% of students with marks below 20% (including a couple who handed a blank paper) and between 25 and 31% of students with marks above 80% (including some with nearly full marks).

The undergraduate statistics course using R was designed to include a combination of the mathematical principles and rationale behind statistical testing, as well as their (computer assisted) practical application. This is done with a combination of approaches, including lectures, problem-solving, and assessment for learning, with a mix of contact time and independent learning, using real datasets and relevant contextual examples. The students have opportunities to practice and act on feedback, and collaborative work and peer support are encouraged and facilitated. However, within these modules the students are still learning statistics in a compartmentalised way, with little transference between modules and across years. As a result, they do not retain much of the information (and need to re-learn it) as they progress between years and they generally still struggle to confidently apply statistical analysis outside the context in which they were taught. This is not unexpected or unique to our students (e.g. [10]), since statistical skills require practice and are often only truly learned later in the learning career, when researchers are finally faced with their own data of interest (e.g. when students need to analyse their own data for their final-year research project). A more integrated approach, making the most of real-life applications with a good support network and more continuous assessment, should greatly improve the students' ability and confidence to apply statistical methods in new contexts and facilitate the teaching and learning of more complex software such as R.

Undergraduates' Responses to Learning Statistics Using R

Figure 5 shows word clouds of the positive and negative comments received, and Table 4 summarises the themes identified across all the feedback comments, with specific examples provided. From 151 feedback comments received by undergraduates being taught statistics through R in the Cardiff School of Biosciences between 2014 and 2016, approximately half (75) had a negative tone, albeit sometimes containing some positive elements. Examples of comments are: "Will be useful, but really confusing and exhausting", or "R hard to use, prefer excel. Understand with time, R will be useful to have". The remaining comments (76) had a positive tone, even if they acknowledged difficult aspects of the learning process, such as: "Initially challenging, less daunting and more enjoyable as I began to work on it", or "difficult but interesting and challenging". A word-cloud analysis of the feedback received showed that among the negative comments, from the total of 47 words that conveyed attitudes towards the learning experience (mostly adjectives), the more commonly used were "difficult" (14), "complicated" (8), "hard" (7), "confusing" (5), and "struggled" (5). The word "time" was also mentioned eight times, always referring to R being time-consuming to learn. Among the positive comments, from a total of 77 words describing such attitudes, the more commonly used words were "enjoyed/enjoying/enjoyable" (18), "fun" (9), "confident" (8),

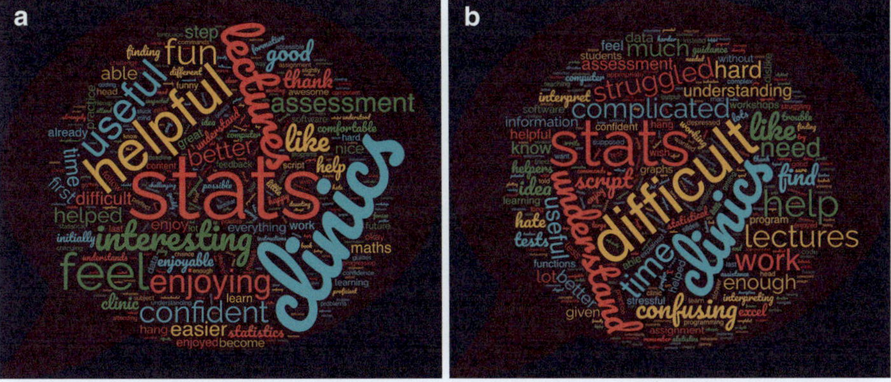

Fig. 5 Word cloud of comments classified as positive (**a**) and negative (**b**), received from undergraduate students from Cardiff School of Biosciences regarding statistics teaching between 2014 and 2017

Table 4 Undergraduate students' comments on their experience of learning data analysis using R statistical software in the Cardiff School of Biosciences across three academic years between 2014 and 2017

Global theme	Description	Examples
R-Software	Sense of discovery and excitement	"Stats is really easy if we know how to read R, which is the changeling part and exciting" "No way I'll be going back to excel now, I'm actually finding this work quite relaxing" "I enjoyed it thoroughly mainly because of R. I was afraid during the lectures that I wouldn't get the hang of it but after the helpful clinics everything was fantastic!" "Enjoy problem solving. New codes exciting. R is great!"
	Not as bad as expected/ not sure at first but won over	"Not as horrible as I thought after starting it properly" "Still not able to manage R but deeply motivated to learn how to do so. It will be a useful tool in future studies" "Sceptical at first but completing the assessment has encouraged me really happy to be learning the skill" "I used to hate stats but I am enjoying this challenge and the fact that I can use a computer for stats, not just pen and paper. It's changing my mind about stats now"
	Time-consuming and difficult to learn—needed more time and teaching	"R fun, but stressful as a coursework task. May enjoy it more if I had allowed more time" "The R software if very unfriendly and complicated and took forever to try to learn to use" "R complicated - need more in-depth lectures. Unsure how to interpret some tests"
	Struggling and finding it demotivating	"Figuring out R stressful; thrown into the deep end" "Struggling. Found R very hard to understand and use. Not sure I'll be able to use its potential. Seems a very complicated way of doing stats" "Using a complex programme like R put me out of my comfort zone; really struggled"
Teaching and support	Importance of clinics	"The stats clinics were the best help, I wasn't confident with stats and R, but they were very helpful" "Stats clinics helped a lot" "I understood a lot more about R when attending the clinics"
	Needed more support	"Stats clinics great, but not enough helpers" "Not many teachers in stats clinic, barely got any one to one help because everyone need help. Struggled a lot" "I wish there would be more lectures/clinics where we were able to be taught R step by step from zero"
	Friendly teaching and learning environment	"I enjoy the clinics as the people there are very helpful and friendly" "Stats clinics were very helpful, people were very nice" "The stats clinics are very useful! The PhD students that help are really helpful and its nice to be able to work on stats with a number of other people around"
	Sense of humour and enjoyment during teaching and learning	"The stats lectures are funny but R takes practise. The statistical data is interesting though" "Lectures are very enjoyable and the module is being taught with a good sense of humour, which is really appreciated" "Really like the format/style of the lectures. They're made quite fun, I just don't like the content/topics/MATHS"

"interesting" (7), and "good" (7). Of the feedback comments with a negative tone, 41% (i.e. 20.5% of the total comments) expressed deep negative feelings towards learning R, conveyed in words like "hate", "anxious", or "struggle"; and 26% of the comments with a positive tone (i.e. 13% of the total) expressed deep positive feelings towards using R, conveyed in words like "enjoy", "awesome", or "brilliant". Considering that statistics is notoriously perceived as a difficult topic, and that students were learning a command-based statistical software (i.e. a new language) while learning statistics, these comments seem encouraging and a testimony to the proposal that teaching statistics using R can be an effective option even at undergraduate level.

2.3   Postgraduate Teaching

Our set-piece teaching is typically presented in a series of 10–12 two-hour sessions, covering at postgraduate level what we consider to be the data analysis methods that most biologists would use most of the time (see Table 3). These sessions comprise a mixture of lecture-style explanation of statistical theory and the rationale for different methods. This is combined with "hands-on" application of this information by analysing one or more real-world datasets of relevance to the course participants and by using template script files that we provide to the students. These sessions are relatively informal compared to a traditional lecture; so interruptions, questions, and discussions are positively encouraged, making the session as much as possible a "conversation" between teachers and learners.

The sessions are typically staffed by the "session leader" plus 2–4 "demonstrators". The session leader presents the taught material by talking through the rationale and implementation of the session topic and they guide the session as a whole. The demonstrators are tasked with facilitating the less formal teaching; this can take a range of forms, including asking their own questions (addressed both to the session leader and to the students), contributing their own views and experiences, and helping the students to implement the worked examples, annotate script files and resolve error messages.

Annotation by Learners of Template Analysis Scripts

Our initial attempts to teach R involved getting students to write their own scripts from scratch, often achieved by students copying lines of script from a PowerPoint display into their own script file. This was problematic because a major hindrance to running scripts in R is the problem of small "typo" errors (e.g. misplaced commas and brackets, case-sensitive errors, etc.) leading to incomprehensible error messages and, consequently, to despairing students. We rapidly realised that a much more effective approach is to provide template script files with generic phrasing, which students can readily edit to apply to their own data. The use of template scripts has proven to be much less frustrating for the students than typing out scripts from scratch and has allowed a greater focus on the *meaning* of the script, the interpretation of the output, and the bigger-picture explanation of what the script *does* rather than being distracted by misplaced commas and error messages.

An important aspect of the use of the template scripts during the lecture-workshop sessions is the annotation of the script by each student, using "hashtag notes" to label each section of script with their own explanations (everything written to the right-hand side of a hashtag # symbol is read by R as a comment, and not implemented as a command). In this way, the basic template scripts that we provide are built up by each student into a bespoke annotated library of scripts, from which students can copy, paste, and adapt to each future analysis that they need to perform.

Guidebook with Generic Script Coding that Maps Onto Our Other Teaching Materials

There are many excellent R textbooks available, which we recommend to our students (Table 5). We encourage our course participants to have at least one favourite textbook that they are especially familiar with for detailed explanations, worked examples, and interpretations of R output.

We have also produced our own bespoke guidebook to accompany our courses [1]. This is a guidebook rather than a textbook in that it is a concise explanation of R-code. It is written in a generic style that maps onto our other teaching materials (PowerPoint slides, template scripts, and worked examples from the lecture-workshops) and provides a unified set of materials that students can refer to beyond the duration of the course itself. The guidebook covers all of the methods that we teach on our courses, plus other methods that we do not necessarily fit into the courses but which we also consider general enough to be included. This guidebook represents a quick-reference resource that students can refer to whenever they need to implement a particular method. From the guidebook, they can obtain the relevant generic scripts that they can then adapt to the dataset that they need to analyse, without needing to find relevant scripts from among lengthy worked examples, as is typically the case with textbooks.

The printed guidebook has been well received by the students that we have taught on our courses, as well as by thousands of R-users from around the world who have purchased the guidebook by mail-order (via www.eco-explore.co.uk) or accessed the Kindle edition "Teaching Biostatistics in Medicine and Allied Health Sciences" (via www.amazon.com).

Table 5 Textbooks that we recommend to our postgraduate students during our courses on data analysis with R statistical software

Authors	Date	Title	Publisher
Mick Crawley	2012	The R Book (2nd edition)	Wiley
Calvin Dytham	2010	Choosing and using statistics	Wiley-Blackwell
Andy Field	2012	Discovering Statistics Using R	Sage Publications Ltd.
Alan Zuur et al.	2009	Mixed Effects Models and Extensions in Ecology with R	Springer-Verlag New York

Informal, Learner-Led "Data Analysis Clinics"

For several years, we held weekly 3-hour drop-in data analysis "clinics" that were open to all bioscience undergraduates, postgraduates, post-doctoral researchers, and academic staff. Anyone could come along with any sort of query, no matter how small or large, and sit down with us for a one-to-one discussion of the problem. We particularly encouraged learners to bring along their R-scripts and data files, so that we could work with them on refining and annotating their scripts. Many of the people who attended these clinics were regular attendees who came to work on a whole analysis over several weeks, in a supportive and collaborative environment, while others visited only occasionally to ask specific questions. The clinics aimed to improve the student learning experience, encourage detailed conversations about data analysis, promote research-driven teaching, and facilitate high-quality data analysis among staff and students.

The informal drop-in format enables us as statistics teaching staff to work more efficiently by dealing with queries in a single block of time, rather than in a series of individual consultations scattered throughout the week. In this way, we can deal with numerous queries in an efficient manner, and yet tailor our teaching to individual learners with a diversity of learning styles and educational needs. We view these clinics as being an important (and popular) element of supporting R-learners through the transition to independent data analysis.

Friendly Online Support via Email

In addition to our weekly drop-in clinics, we received a large number of email queries asking for help with data analysis in general and with R in particular. The volume of email traffic can be problematic; indeed, this is why we initiated the in-person clinics so that many such queries could be dealt with more efficiently in a shorter amount of time. Nevertheless, it is often necessary to respond by email (for a variety of reasons), and again our approach is to be friendly and encouraging to learners who may be struggling with the seemingly user-unfriendly R interface, or bemused by incomprehensible warning messages.

"R-Space": A Facebook Forum that Facilitates Peer-to-Peer Interaction as Well as Expert Input

We encourage our students to be active members of the worldwide network of R-users. One way in which we do this is through "R-Space", which is our Facebook forum for discussions, dissemination of news, ideas, and requests for help with data analysis in R. Students are used to interacting on social media, including Facebook, and so interactions are relatively informal. We maintain an ethos of "there are no stupid questions", in that if it is a relevant question for one person, then it is likely to be a relevant question for lots of other people. This intentionally friendly and open approach is a deliberate alternative to the rather stern and forbidding tone of many of the other R discussion forums (Table 1), where questions from beginners are often met with a dispiritingly harsh and off-putting response. The deliberately welcoming ethos of R-space aims to encourage participation by beginners as well as more experienced R-users and to facilitate peer-to-peer teaching as learners begin to feel confident in sharing their own advice, insights, and experience with others. R-space currently (2022) has around 3400 participants from around the world, many of whom have never been on one of our courses, but who enjoy learning new things for themselves and teaching others on the forum.

Video Tutorials Enabling Independent Learning

We have produced a series of video tutorials on common data analysis methods, using the same content and analysis scripts as in our other teaching materials, which students can download and view independently. These videos can be used by students to recap on the rationale for different methods and to see a worked example being implemented, discussed, and reported both visually (as graphs) and quantitatively. The main advantage of this approach is that students can pause or replay the videos while they run the scripts on their own data, which can provide a sense of learning at their own pace. At best, these tutorials could enable students to learn fully independently, but they can also help students to understand the background and general principles of a method before we work on implementing the analysis together during the drop-in clinics.

Postgraduates' Responses to Learning Statistics Using R

Postgraduate students come to our data analysis courses with a great diversity of different backgrounds in learning and applying statistics with different software. For this reason, we assume no prior knowledge and aim to begin with very simple elements that build up quite rapidly across the course programme to reach a considerable degree of complexity. Figure 6 shows a word cloud of the comments received by postgraduates attending the courses and a selection of specific feedback on their experience of these courses is shown in Table 6.

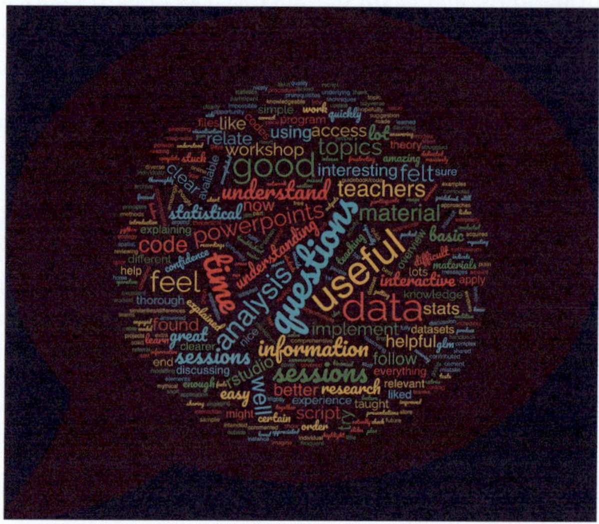

Fig. 6 Word cloud from comments received from postgraduate researchers who attended Exo-explore R courses

Table 6 Postgraduate students' comments on their experience of learning data analysis using R statistical software in the Cardiff School of Biosciences across three academic years between 2014 and 2017

Global theme	Description	Examples
R-software course	Range of topics covered	"Just amazing ambition to cover everything and well accomplished, but may be good to indicate the prerequisites to the course, such as some basic statistics knowledge or so?" "Really thorough and interesting course. The course leaders were dedicated to each participant and were really engaging. Thank you so much!" "A very good introduction to a wide range of analysis approaches (….)"
	Complex content	"(…) the PowerPoint slides I felt were slightly too focused on how to implement it in R and not on understanding the underlying stats"
Teaching and support	Enthusiasm and availability/ individual support	"With enthusiastic and approachable instructors, an excellent guidebook and a workflow for approaching all sorts of analyses, you can't go wrong" "The course leaders were dedicated to each participant and were really engaging" "(…) taught with a lot of enthusiasm (which really contributed to the excellent quality of the course)"
	Opportunity for questions, discussion, and practice	"I very much appreciated the time teachers gave to answering questions and discussing topic further during breaks and after the teaching day, great stuff!" "Good pace and lots of opportunities to ask questions, lots of useful worked examples as well as covering the theory behind the tests" "A very thorough overview of different statistical analysis techniques, with opportunity to ask frequent questions and relate to your own research" "This course was by far the best I have ever attended! Great student/teacher ratio, [with] plenty of time to apply the newly learnt knowledge to your own data"

3   Conclusions

In this paper, we have set out our own experience-led approach to teaching statistics to life scientists using R. Although the command-line format presents a steep learning curve and dependence on programming, R provides an important opportunity for deeper understanding of data analysis that is less readily taught with menu-driven programmes. Our research-led teaching provides learners with a unified platform for data analysis that develops across their undergraduate and postgraduate career. Ultimately, this provides learners with the tools to perform a wide range of the most commonly used analyses and a logical framework to expand this knowledge to analyse any sort of dataset that they may encounter in the future as independent scientists. Furthermore, the archiving of datasets with annotated analysis scripts represents an important facility for making a permanent record of workflow, improving the transparency and replicability of data analysis. Importantly, as R is free, students can readily continue to use and update R after they graduate, if they move to another institution or take up posts with non-university organisations.

Overall, R is an outstanding resource for teaching and learning about statistics and for creating transparent and replicable analyses and publication-quality data visualisations. Learners who gain confidence and competence in using R are gaining valuable skills that will benefit them for the rest of their scientific careers. This said, it is likely that only a small proportion of undergraduates in life sciences will engage in scientific careers that involve statistical analysis, but this could be argued for many other skills they learn during their undergraduate degree. For those who do, learning R can be highly advantageous and for those who do not, a critical understanding of statistics is useful in a wide range of contexts and familiarity with programming skills is growing in relevance with some evidence pointing to many associated cognitive benefits that are highly transferable [29].

Our reflections are from teaching large cohorts of biosciences student from diverse backgrounds and a wide range of career aspirations. Hence, these insights should be directly relevant to all life sciences (and beyond), including medical or dental undergraduates and postgraduates. The statistics clinics (or a similar forum in which the students can work on tasks independently but with support at hand) seemed key for successful teaching statistics using R. This requires students' and staff's time and can be a limiting factor if statistics are only a very small part of the curriculum, as it is often the case in medical teaching contexts.

Acknowledgements We thank the many long-suffering students who have—whether they intended to or not—helped to improve our teaching of statistics using R over recent years.

References

1. Thomas RJ, The Guidebook Team. Data analysis with R statistical software: a guidebook for scientists. Caerphilly: Eco-explore; 2017.
2. Labov JB, Reid AH, Yamamoto KR. Integrated biology and undergraduate science education: a new biology education for the twenty-first century. CBE Life Sci Educ. 2010;9:10–6.
3. QAA. Subject benchmark statement - biosciences. UK Quality Code for Higher Education. 2019. Available https://www.qaa.ac.uk/docs/qaa/subject-benchmark-statements/subject-benchmark-statement-biosciences.pdf. Accessed 12 January 2022.
4. General Dental Council Outcomes for Graduates. 2015. Available https://www.gdc-uk.org/docs/default-source/quality-assurance/preparing-for-practice-(revised-2015).pdf?sfvrsn=81d58c49_2. Accessed 12 January 2022.
5. General Medical Council Outcomes for Graduates. 2018. Available https://www.gmc-uk.org/education/standards-guidance-and-curricula/standards-and-outcomes/outcomes-for-graduates/outcomes-for-graduates. Accessed 12 January 2022.
6. A'Brook R, Weyers JDB. Teaching of statistics to UK undergraduate biology students in 1995. J Biol Educ. 1996;30(4):281–8.

7. Cavenagh P, Leinster SJ, Miles S. The changing face of medical education. Milton Keynes: Radcliffe Publishing; 2011.
8. Gal I, Ginsburg L, Schau C. Monitoring attitudes and beliefs in statistics education. In: Gal I, Garfield JB, editors. The assessment challenge in statistics education. Amsterdam: IOS Press; 1997. p. 37–51.
9. Swift L, Miles S, Price GM, Shepstone L, Leinster S. Do doctors need statistics? Doctors' use of and attitudes to probability and statistics. Stat Mcd. 2009;28(15):1969–81.
10. Colon-Berlingeri M, Burrowes PA. Teaching biology through statistics: application of statistical methods in genetics and zoology courses. CBE Life Sci Educ. 2011;10:259–67.
11. Medeiros R, Thomas RJ. Learning statistics - the wonders and hindrances of statistical software. Manuscript in preparation. 2022.
12. Hackenberger BK. R software: unfriendly but probably the best. Croat Med J. 2020;61(1):66–8. https://doi.org/10.3325/cmj.2020.61.66.
13. Muenchen RA. The popularity of data science software. 2010. Available https://r4stats.com/articles/popularity/. Accessed 12 January 2022.
14. Shuker DM. Review of getting started with R: an introduction for biologists, data analysis with R statistical software. A guidebook for scientists and discovering statistics using R. Anim Behav. 2012;84(6):1597–600. https://doi.org/10.1016/j.anbehav.2012.09.019.
15. Ihaka R, Gentleman R. R: a language for data analysis and graphics. J Comput Graph Stat. 1996;5:299–314.
16. Tippmann S. Programming tools: adventures with R. Nat News. 2015;517(7532):109.
17. Pollack RD, Klimberg RK, Boklage SH. An examination of the open-source software package R. ORMS-Today. 2015;42:5.
18. TIOBE Index. TIOBE – The Software Quality Company. 2021. https://www.tiobe.com/tiobe-index/. Accessed 16 August 2021.
19. JASP Team. JASP (version 0.16). 2021. https://jasp-stats.org/. Accessed 12 January 2022.
20. The Jamovi Project. jamovi (version 1.6). 2021. https://www.jamovi.org. Accessed 12 January 2022.
21. Selker R, Love J, Dropmann D, Moreno V. jmv: the 'jamovi' analyses. R package version 2.0. 2021. Accessed 12 January 2022.
22. Aikens ML, Dolan EL. Teaching quantitative biology: goals, assessments, and resources. Mol Biol Cell. 2014;25(22):3478–81.
23. GAISE College Report ASA Revision Committee. Guidelines for assessment and instruction in statistics education college report 2016. 2016. Available http://www.amstat.org/education/gaise. Accessed 12 January 2022.
24. Basturk R. The effectiveness of computer-assisted instruction in teaching introductory statistics. Educ Technol Soc. 2005;8(2):170–8.
25. Cobb G. Teaching statistics. In: Heeding the call for change: suggestions for curricular action, vol. 22. Washington: Mathematical Association of America; 1992. p. 3–43.
26. Bialek W, Botstein D. Introductory science and mathematics education for 21st-Century biologists. Science. 2004;303(5659):788–90.
27. Mustafa RY. The challenge of teaching statistics to non-specialists. J Stat Educ. 1996;4:1. https://doi.org/10.1080/10691898.1996.11910504.
28. Sowey ER. Teaching statistics: making it memorable. J Stat Educ. 1995;3:2. https://doi.org/10.1080/10691898.1995.11910487.
29. Scherer R, Siddiq F, Sánchez-Scherer B. Some evidence on the cognitive benefits of learning to code. Front Psychol. 2021;12:559424. https://doi.org/10.3389/fpsyg.2021.559424. PMID: 34566735; PMCID: PMC8458729.

Statistics in a World Without Science

Christopher R. Tench

1 Introduction

Statistics is important in medical research but can easily be used incorrectly, and unfortunately the evidence suggests this is common. There is now a recognized replication crisis [1–4], and replication of effect is always needed (if possible) for conclusive science. It is understood that when sampling and statistical inference are employed there are false conclusions from type 1 and type 2 errors. However, lack of replicability appears to be at a higher rate than any reasonable and skilled scientist might purposefully design into their experiments. Indeed, some attempts to reproduce the results of medical research have failed more times than they have succeeded [3, 5]. Here, the type 1 and type 2 errors are not the major concern, as they are to be expected and are random. Instead, the focus is on avoidable systematic bias caused by lack of adherence to good scientific practice, which must come as no surprise as the philosophy of science is not generally taught. Without resorting to teaching philosophy, which is probably not appropriate for clinicians in general, how can the replication crisis be addressed?

The understood requirement for testing scientific hypotheses and acquired knowledge of statistical null hypothesis significance testing (NHST), without the understanding that the two are not similar [6], explains where much of this 'crisis' occurs. The eager researcher wanting to test their hypotheses and armed with knowledge of NHST may walk straight into the common trap of assuming they can do the former by performing the latter, yet no statistical testing approach, frequentist or Bayesian, can directly perform scientific testing. Inappropriate use of statistical

C. R. Tench (✉)
Mental Health and Clinical Neurosciences, Clinical Neurology, University of Nottingham, Queen's Medical Centre, Nottingham, UK

NIHR Nottingham Biomedical Research Centre, Queen's Medical Centre, University of Nottingham, Nottingham, UK
e-mail: christopher.tench@nottingham.ac.uk

© The Author(s), under exclusive license to Springer Nature Switzerland AG 2023
D. J. J. Farnell, R. Medeiros Mirra (eds.), *Teaching Biostatistics in Medicine and Allied Health Sciences*, https://doi.org/10.1007/978-3-031-26010-0_13

inference undermines the generally accepted scientific requirement for falsifiability [7]; falsifiability has its critics [7], but is likely to be part of the assumed scientific framework employed unless another framework is specified. Unplanned exploratory analysis combined with flexible statistical models produces statistically significant 'discoveries' that often do not stand up to attempts at replication [8, 9]. Such discoveries are relatively easy to find by determined exploration, so the frequency with which the researcher is forced to conclude that their hypothesis has been falsified is low; presumably there are very few scientists that are genuinely never wrong! This leads to such biased practice as HARKing (hypothesizing after the results are known) [10], where researchers convinced retrospectively of the validity of their discoveries, then imply that testing them was always the intension. There are some strong motivational factors behind this, including studies without statistically significant results being relatively difficult to publish [11]; a problem that may be getting worse [12]. However, it is also likely that a lack of awareness is a major cause of unscientific practice since the growing literature covering the replication crisis is somewhat specialist for the non-scientist or non-statistician.

With philosophy of science courses being uncommon and possibly unsuitable, the inevitable push to reduce the impact of the replication crisis may fall squarely at the feet of those teaching statistics [13]; inevitable because continuing experiments that too often cannot be reproduced on human or animal subjects is surely not an option. It should be noted that what is written here is well established. Indeed Jacob Cohen, who details the problems in 1994 [14], emphasizes that this was old news even then. This manuscript is therefore just another, of many, reminders that we have a problem. Some suggested fixes have been reported, such as pre-registration [15] or the banning of statistical significance [16]. Whenever, and however, the required change does come, teachers of statistics will likely be the ones faced with the challenge of how to adapt to it and help the 'move to a world beyond $p <$ 0.05' [17].

1.1   Problems

It would not be uncommon for researchers to use statistical null hypothesis significance testing beyond its intended use. Belief that NHST can be used as an exploratory tool and provide scientifically conclusive results is apparent on reading clinical research papers. This has been expressed clearly by Richard Horton [9], editor of the Lancet, who says we have invalid exploratory analyses and a love of 'significance' polluting the literature with statistical fairy-tale. This is not just opinion, as the p-value makes no appeal to the scientific method or the scientific hypothesis under test. The act of fitting a statistical model to data is an optimization (least squares or maximum likelihood, for example) to provide the best fitting model, but best is not equal to scientific validity. The only way to infer scientific meaning by the use of NHST is to purposefully design it into an appropriate experiment such as to give it meaning. But much of research involves unplanned statistical analyses, where the requirements of using NHST to identify scientifically conclusive results may not be met. Indeed, without access to a protocol that was followed and clearly produced before there was access to the data, there is no way to know what analysis was actually performed. Where only the results the author decided to present are made available, the hypotheses may simply have been retrofitted to the data, providing no conclusive evidence without an independent validation. This particularly unfortunate practice means those studies that are well designed and planned may be put at a disadvantage compared to those where flexible analysis and retrofitting of hypotheses is conducted and undisclosed on the basis of the misunderstanding that NHST lends the work some measure of scientific validity.

Design of experiments may be challenging to clinicians that probably do not have the time to invest in advanced statistics, so the need for access to statistical expertise is clear. However, it is certainly possible to warn of the pitfalls of exploratory NHST. Avoiding these pitfalls may reduce the incidence of unscientific conclusion in the literature and reduce the risk when attempting to validate findings, or when basing research decisions on published claims.

Common Pitfalls of Exploratory Analysis Using NHST

Statistical Hypothesis Testing Does Not Test a Scientific Hypothesis

Perhaps the most fundamental issue is the mistaken belief that statistical hypothesis testing can test scientific hypotheses. This is somewhat understandable, after all the words 'hypothesis testing' is used to describe the topic. However, an unbiased and controlled experiment is needed in order to draw any scientifically valid conclusion about the hypothesis under test. Statistical hypothesis testing, either Bayesian or frequentist, that uses experimental data but does not meet these requirements does not constitute a valid scientific test of a hypothesis. A scientific test requires a proposed (hypothesized) effect to test, and the use of random sampling necessitates a variance on that effect in order that a suitable experiment might be designed with understood error rates to determine the strength of evidence (rate of false conclusion) it can produce. Conversely, NHST has no concept of a proposed scientific hypothesis, producing results based only on an assumed null hypothesis. As an example, consider the typical hypotheses involved in a t-test $\mu = \mu_0$ vs $\mu \neq \mu_0$. The inability to quantify the type 2 error rate for the alternative statistical hypothesis $\mu \neq \mu_0$ means the t-test is not falsifiable using random sampling. Furthermore, a significant result offers no evidence in favour of any scientific hypothesis as it is explicitly assumed the statistical null is true.

Data-Dependent Analyses Are Biased

Data-dependent analyses occur when decisions about the analysis are not made a priori but are left until the data are available to guide those decisions. Acquiring experimental random samples generally demands a prospective decision on what data to collect and why; an exception to this is the use of data collected and made available to researchers, for example, the UK biobank [18], where those decisions are made independently of the study plans. Often, however, a decision on what to do with the sample is delayed until the analysis point; indeed, it is this point where advice is often sought from the statistician, which according to a quote attributed to Fisher [19] is asking for a postmortem to find out how the experiment died. Looking at the sample, seeing trends and subsequently making decisions about which statistical models to fit or analyses to perform is commonly the course of action, resulting in researcher degrees of freedom [8] that can easily lead to a high rate of false conclusions. Such data-dependent analysis decisions produce p-values that are difficult to interpret because the analysis performed is inevitably biased towards one particular random sample from all possible random samples.

No Appropriate Consideration of the Sample Size Means No Appreciation of Error Rates

It is very common that samples are small so power may be low and the error rate high [20]. This is particularly so if multiple statistical tests are to be conducted. Correction for multiple tests is often considered only in terms of the type 1 error rates (Bonferroni for example) only, which reduces power further. While it may be understood that low statistical power can cause the researcher to miss true effects, what may not be so well appreciated is that it also increases the probability that a significant result is a false positive [20]. Without consideration of sample size and power, no estimate of the error rate in results is available, so conclusions cannot be conclusive.

The Experimental Null and Statistical Null May Be Different

The statistical null is defined and understood. However, the common practice of performing multiple data-dependent analyses makes it difficult to rigorously control experiments; experimental control is part of experimental design and requires the analyses to be planned or risk missing important controlling factors. Without experimental control, the assumed statistical null may not be correct. This is recognized by Cohen who refers to the statistical null as the nil hypothesis, which is always false [14]. The issue may not be mitigated with large samples since any systematic lack of control (because research is ongoing and not every important controlling factor might be known) may then produce significance; Cohen [14] provides a quote suggesting that under such circumstances hypothesis testing on large samples is simply trying to confirm the sample is large, which the researcher already knows. Without careful consideration of the controlled experiment, the conclusions based on rejecting the assumed statistical null hypothesis can easily be due to systematic bias. It should be recognized that randomized control trials are an exception to this since the randomization of one population creates a controlled experiment.

A Good Fitting Statistical Model Does Not Imply Validity

It is a common procedure to build statistical models that explain the sample data and then infer confidence in the model by referring to statistical significance or by retrospectively citing supporting literature. Building statistical models is a process of selecting one from a possible vast (consider the number of possible combinations in a regression model containing multiple independent variables and interactions) range of models and it is common that one or more may fit well, but potentially only because of model flexibility [21, 22]. The selection procedure often involves searching for the optimal fitting model, which generally means minimizing the difference between the data and the model and is in no way constrained by ability to explain independent future data. Models built in this way cannot be considered scientifically valid simply because of apparent statistical significance, or because they 'make sense' retrospectively. Model building is necessary, but conclusions drawn must recognize the need for further validation.

1.2   Simple Solutions

Students of statistics need to be aware of the problems of invalid exploratory analysis both for their own studies and for identifying the limitations of published clinical research. Addressing the replication crisis has already been considered extensively. The suggestions include pre-registration and avoiding statistical significance, unless correctly designed into a proposed study [23].

Study protocols can be made available either as pre-registered studies or by making a document available such that it is citable using a digital object identifier. Pre-registration offers the advantage that citations can be accumulated while the study is being conducted. The major advantage of producing, and adhering to, a protocol is avoiding data-dependent analysis. Instead, the analysis plan is determined only using the knowledge of the researcher and their scientific experience.

Avoiding unjustified NHST (the p-value fallacy [24, 25]) eliminates the general bias that those results with $p < 0.05$ (usually) are considered of greater importance than those with $p > 0.05$ even when there is nothing in the experimental design to make this distinction meaningful. An alternative is simple estimates of effect [14]. These can then be discussed by reference to details and expectations expressed in the study protocol. Of course, estimates and their confidence intervals, or credible intervals for the non-frequentist, are still prone to the issues of bias and multiple testing, but there is at least less tendency to the false dichotomy from unplanned NHST.

More advanced alternatives to p-values, that may reduce the possibility of producing false results that would then vanish upon attempting possibly expensive replication work, is to evaluate the quality of fitted models on hold-out data using cross-validation [26]. This provides a hint of the ability of the model to predict unseen data and is likely to be incorporated into more studies as machine learning becomes more accessible in medical research. As always, this method is not a solution to unplanned exploratory analyses, which may negate any advantage gained.

1.3   What About Discovery?

It is often claimed that exploratory analysis is in the name of discovery, and this is of course part of science. Exploration can help generate hypotheses that may later be tested. However, it must be understood that the meaning of the p-values is not defined under such experimental conditions for the reasons listed above. Furthermore, for every hypothesis generated with the help of exploratory data analysis, there must be a subsequent, possibly expensive, independent test before any scientific conclusion can be drawn. It is good practice, and common sense, to therefore take steps to minimize the false discoveries.

Exploration commonly involves some element of data-dependent analysis, which risks high rates of false 'discoveries' or false negatives if based on NHST. Inevitably there might be findings that are only a result of random sampling, bias, or inappropriate statistical analysis. A prospective plan based on current knowledge and logically expected outcomes makes it possible to avoid some data-dependent analysis. The researcher can always say something before the experiment, after all the choice of data to collect was made a priori and presumably with reasons. This is better than performing a literature search for supporting evidence retrospectively, particularly with the replication crisis impacting that literature. However, it might be argued that such self-imposed restriction could lead to missing some important 'discovery', but the researcher could even prospectively plan to conduct exploratory analyses providing they also plan to conclude that scientific evidence for findings require further independent research.

2   Conclusions

To draw scientific conclusions using statistical inference, an avoidance of bias and awareness of false conclusion rates is nonoptional. Any statistical analysis should be part of the study plan/design or risk inflating these. Only with such a plan is it possible to know that all analyses were performed as planned, removing the possibility of HARKing or selective reporting, and that extra unplanned analyses are identifiable so they may be interpreted with appropriate caution. Even in the absence of a testable hypothesis, data analysis can always be planned and justified prospectively if the researcher is an expert. It is even possible to plan to do data-dependent analyses providing there is also a plan to conclude only that any findings require further independent study before any scientific conclusion can be drawn. Producing, and adhering to, a study protocol also has added advantage of providing evidence of good scientific practice. Similarly, avoiding statistical significance except where the meaning is designed into the experiment will prevent the invalid hunt for $p < 0.05$. Students can be informed of the need for such a prospective approach and appropriate conclusions without resorting to specialist training in the scientific method. More sophisticated approaches such as cross-validation might also be helpful but may require advanced training.

References

1. Reproducibility and reliability of biomedical research. The Academy of Medical Sciences [Internet]. [cited 2021 Oct 7]. Available from https://acmedsci.ac.uk/policy/policy-projects/reproducibility-and-reliability-of-biomedical-research.

2. Stupple A, Singerman D, Celi LA. The reproducibility crisis in the age of digital medicine. NPJ Digit Med. 2019;2(1):1–3.

3. Baker M. 1,500 scientists lift the lid on reproducibility [Internet]. Nature 2016 [cited 2021 Oct 8]. Available from https://www.nature.com/articles/533452a.

4. Amrhein V, Korner-Nievergelt F, Roth T. The earth is flat (p > 0.05): significance thresholds and the crisis of unreplicable research. PeerJ. 2017;5:e3544.

5. Open Science Collaboration. Estimating the reproducibility of psychological science. Science. 2015;349(6251):aac4716.

6. Dixon P, O'reilly T. Scientific versus statistical inference. Can J Exp Psychol. 1999;53(2):133–49.

7. Mitra S. An analysis of the falsification criterion of Karl Popper: a critical review. Tattva J Philos. 2020;12(1):1–18.

8. Simmons JP, Nelson LD, Simonsohn U. False-positive psychology: undisclosed flexibility in data collection and analysis allows presenting anything as significant. Psychol Sci. 2011;22(11):1359–66.

9. Horton R. Offline: what is medicine's 5 sigma? Lancet. 2015;385(9976):1380.

10. Kerr NL. HARKing: hypothesizing after the results are known. Personal Soc Psychol Rev. 1998;2(3):196–217.

11. Dickersin K, Chan S, Chalmersx TC, Sacks HS, Smith H. Publication bias and clinical trials. Control Clin Trials. 1987;8(4):343–53.

12. Joober R, Schmitz N, Annable L, Boksa P. Publication bias: what are the challenges and can they be overcome? J Psychiatry Neurosci. 2012;37(3):149–52.

13. Perezgonzalez JD. Fisher, Neyman-Pearson or NHST? A tutorial for teaching data testing. Front Psychol. 2015;6:223.

14. Cohen J. The earth is round (p < .05). Am Psychol. 1994;49:997–1003.

15. Nosek BA, Ebersole CR, DeHaven AC, Mellor DT. The preregistration revolution. Proc Natl Acad Sci. 2018;115(11):2600–6.

16. Amrhein V, Greenland S, McShane B. Scientists rise up against statistical significance. Nature. 2019;567(7748):305–7.

17. Wasserstein RL, Schirm AL, Lazar NA. Moving to a World Beyond "$p < 0.05$". Am Stat. 2019;73(sup1):1–19.

18. Sudlow C, Gallacher J, Allen N, Beral V, Burton P, Danesh J, et al. UK Biobank: an open access resource for identifying the causes of a wide range of complex diseases of middle and old age. PLoS Med. 2015;12(3):e1001779.

19. Kim J, Bang H. Three common misuses of P values. Dent Hypotheses. 2016;7(3):73–80.

20. Button KS, Ioannidis JP, Mokrysz C, Nosek BA, Flint J, Robinson ES, et al. Power failure: why small sample size undermines the reliability of neuroscience. Nat Rev Neurosci. 2013;14(5):365–76.

21. Rencher AC, Pun FC. Inflation of R2 in best subset regression. Technometrics. 1980;22(1):49–53.

22. Hawkins DM. The problem of overfitting. J Chem Inf Comput Sci. 2004;44(1):1–12.

23. Lakens D. The practical alternative to the p value is the correctly used p value. Perspect Psychol Sci J Assoc Psychol Sci. 2021;16(3):639–48.

24. Dixon P. The p-value fallacy and how to avoid it. Can J Exp Psychol. 2003;57(3):189–202.

25. Goodman SN. Toward evidence-based medical statistics. 1: the p value fallacy. Ann Intern Med. 1999;130(12):995–1004.

26. Koul A, Becchio C, Cavallo A. Cross-validation approaches for replicability in psychology. Front Psychol. 2018;9:1117.

Killing Me Softly with Your Stats Teaching: How Much Stats Is Too Much Stats?

Renata Medeiros Mirra and Robert J. Thomas

1 Introduction

This paper reflects on a round-table session at the 2021 Annual Meeting for Teachers of Statistics in Medicine and Allied Sciences (commonly referred to as "Burwalls"), where participants discussed what statistical concepts or skills students should learn at undergraduate level in medical or other health science degrees, and what challenges are presented to statistics teachers in these contexts. Approximately 40 participants joined the discussion at some point and approximately 20 of those were engaged throughout and participated in the task/activities proposed.

2 Setting the Scene

The participants were asked to approach the discussion from their current position as teachers and/or their knowledge of statistics while reflecting on their time as students — in other words, putting themselves in the shoes of students or the people we are currently trying to teach, without letting go of their current position of accumulated knowledge.

We started by reflecting on general ideas about teaching and learning and the contexts in which people learn best. Common sense, backed by literature available on the topic, suggests that the key to effective learning is motivation, and such motivation develops when three elements come together: the topics feel interesting, attainable, and useful [1]. This immediately poses challenges when it comes to

R. Medeiros Mirra (✉)
Cardiff School of Dentistry, Cardiff University, Cardiff, UK
e-mail: medeirosmirrarj@cardiff.ac.uk

R. J. Thomas
Cardiff School of Biosciences, Cardiff University, Cardiff, UK

D. J. J. Farnell, R. Medeiros Mirra (eds.), *Teaching Biostatistics in Medicine and Allied Health Sciences*, https://doi.org/10.1007/978-3-031-26010-0_14

teaching statistics, particularly in medical or health care contexts, where students may feel that they are learning something unrelated to what interested them when they chose their degree; they usually do not perceive statistics as useful or relevant to their future careers, neither do they find it interesting and, despite being generally high-achieving students, many do not feel they are able to learn it either. This presents statistics teachers with an almost impossible combination of challenges, which can be very dispiriting but when overcome can also feel profoundly rewarding; examples of students who go from hating statistics to becoming enthusiastically "nerdy" about it might be rare but are very inspiring. More encouragingly, recent research that focuses on the relationship between teacher and students shows strong links between this relationship and the students' motivation, implying that the relationship between teacher and students is itself at the core of the learning process [2–4].

Literature specifically about teaching and learning statistics is scarce, but some research points to the need for authenticity, for example, when students learn through investigating real-life problems and datasets [5–7], applying the concepts through computer assisted teaching [8, 9], and focusing on understanding of core concepts and underlying principles of statistical methods, instead of mastering specific statistical techniques [9, 10].

Statistics is a core subject in many life-sciences degrees in the UK, including most medical and dental degrees. It is an essential skill for those pursuing careers as scientific researchers and some understanding of statistics is important to any evidence-based healthcare practice. Nevertheless, there is little guidance regarding statistics teaching, both in terms of format (how to teach) and, perhaps more surprising, in terms of content (what to teach). There are no agreed guidelines on what should constitute a good statistics program for degrees in life sciences in the UK. For example, the guidelines from the General Medical Council [11] are mostly vague and only hint at the idea that students should be able to describe the role of qualitative and quantitative approaches and that they should be able to interpret common statistical tests, but they do not specify the core concepts that they need to learn, nor how they should be taught. For dental practice, the General Dental Council [12] guidelines are even less clear and only refer to students being able to make evidence-based decisions and critically appraise scientific research, with three mentions relating to how statistics should be taught scattered across the document.

This leaves statistics teachers with little formal guidance when it comes to planning our teaching programmes. Furthermore, there is rarely a team of statistics teachers for each programme and so individual teachers tend to work in isolation. This can be daunting but also liberating in that there is great freedom and flexibility for those keen on innovating their teaching, but it raises the question of whether we are teaching our students what they need to know. Miles et al. [13] asked established clinical doctors to consider their career experiences and, in retrospect, think of what statistical concepts or methods they thought they should have been taught as undergraduates. The results indicated that clinicians wish their teaching had focus on an "understanding of basic concepts", "a basic grounding", or "core principles". In terms of specific statistical methods, there seems to be some agreement on keeping teaching relatively simple, with respondents talking of "basic data analysis", "simple tests like t-tests", "common tests", and "confidence intervals".

The first task that we set in this round-table session was for participants to complete a survey telling us about their current teaching programmes. Twenty participants engaged with the task, nine of whom taught undergraduates, 10 taught postgraduates, and one taught both cohorts. Table 1 presents all the concepts reported by the participants, split by frequency of inclusion in their teaching programmes. 36.8% of participants reported teaching statistical concepts theoretically, without mathematical basis, 26.3% reported teaching theoretically with mathematical basis, and 63.1% reported teaching the concepts with practical application using either manual calculations or statistical software. None of the participants reported

Table 1 List of concepts taught in undergraduate ($n = 10$) and postgraduate ($n = 10$) programmes in the life sciences in the UK provided by participants of the "Burwalls" 2021 round-table session titled: "Killing me softly with your stats teaching: how much stats is too much stats", reflecting on content and format of statistical teaching in higher education life-sciences courses in the UK

Most taught concepts (>60%)	Intermediately taught concepts (30–60%)	Least taught concepts (<30%)
• P-value (94.7%)	• Paired t-test (57.9%)	• Z-scores (26.3%)
• Confidence intervals (89.5%)	• Variance (57.9%)	• Validity (26.3%)
• Statistical significance (89.5%)	• Sampling bias (52.6%)	• Degrees of freedom (26.3%)
• Types of data (89.5%)	• Mann–Whitney test (52.6%)	• Reproducibility crisis (26.3%)
• Averages (84.2%)	• Effect size (52.6%)	• Kappa statistic (26.3%)
• Standard deviation (84.2%)	• Simple linear regression (52.6%)	• Sum of squares (21.1%)
• Statistical tests (84.2%)	• Critical appraisal (47.4%)	• Kurtosis (15.8%)
• Statistical sample (78.9%)	• Sample size (42.1%)	• Bland–Altman plot (15.8%)
• Clinical significance (78.9%)	• Data skewness (42.1%)	• Survival analysis (15.8%)
• Odds ratios (73.7%)	• Central limit theorem (42.1%)	• History of statistics (15.8%)
• Interquartile range (73.7%)	• Multiple regression/confounders (42.1%)	• Survival analysis (15.8%)
• Normal distribution (73.7%)	• Percentiles/quartiles (42.1%)	• Poisson distribution (10.5%)
• Standard error (68.4%)	• ANOVA (42.1%)	• Binomial distribution (10.5%)
• Study design (68.4%)	• Spearman's correlation (42.1%)	• Mean absolute deviation (10.5%)
• Relative risk (68.4%)	• Wilcoxon signed-rank test (42.1%)	• Normality test (10.5%)
• Type I/type II errors (68.4%)	• Logistic Regression (36.8%)	• Variance test (10.5%)
• Parametric vs. non-parametric tests (68.4%)	• Outliers (36.8%)	• Likert plots (10.5%)
• Histograms (68.4%)	• Pie-charts (36.8%)	• Interval plots (10.5%)
• Boxplots (63.2%)	• Rules of normal distribution (31.6%)	• PCA (5.3%)
• Bar-plots (63.2%)	• Meta-analysis (31.6%)	• Median absolute deviation (5.3%)
• Scatterplots (63.2%)	• Kruskal–Wallis (31.6%)	• Spatial analysis (5%)
• Statistical power (63.2%)		• Reverse causality (5%)
• Independent samples t-test (68.4%)		
• Chi-squared test (63.2%)		
• Pearson's correlation (68.4%)		

teaching Bayesian inference. 26.3% of participants reported having over 40 h of statistics teaching timetabled in the curriculum, 10.5% reported 24 h of teaching timetabled, 31.6% reported between 10–18 h, and the remaining 31.6% reported less than 10 h.

This list also agrees to a great extent with that from the survey of biostatistics teaching that "Burwalls" participants are invited to engage with every year (see chap 1: "A Review of Medical and Dental Statistics Teaching in the UK"), which contains information collected since 2013 compiled on this webpage: https://www.ed.ac.uk/usher/annual-meeting-teachers-of-medical-statistics-2018/overview-of-teaching-of-statistics-within-medicine. It is worth noting that the selection and wording of concepts/topics for people to choose from were not exactly the same on the annual survey and the survey used in this session (e.g., basic concepts like "averages" are not included in the annual survey under the assumption that everyone teaches them).

These results suggest that there is a general agreement between the concepts that most participants reported including in their teaching programmes with those that medical practitioners say they would have liked to have learned as undergraduates. Since students are (and were) being taught the "right content", it is perhaps important to focus on how statistics are taught, as well as what is being taught. This includes considering how long students take to properly assimilate statistical content, even if we are teaching only core principles and basic data analysis skills.

3   Aims of the Session

In the absence of clear guidelines for statistical teaching in the life sciences, the aim of this session was to use the combined experiences and expertise of the participants to produce "Burwalls guidelines" for what an undergraduate curriculum in life sciences should include in terms of statistics teaching, and how those concepts should be approached. In other words, how much stats is enough stats for our undergraduates, and how should we teach it.

4   Process

To facilitate this discussion, we created five different scenarios in terms of time available for formal statistics teaching in the curriculum, going from unlimited time and resources to one single session (Fig. 1). We split the participants into breakout rooms with the different scenarios, asking them what they would teach undergraduates if they had that allocation of time. They were also asked to consider whether they would teach interpretation or application of concepts, with or without software, with or without assessment and in bulk or spreading the teaching across the time available. Each group had 20 minutes to discuss and populate a shared Google document with their ideas.

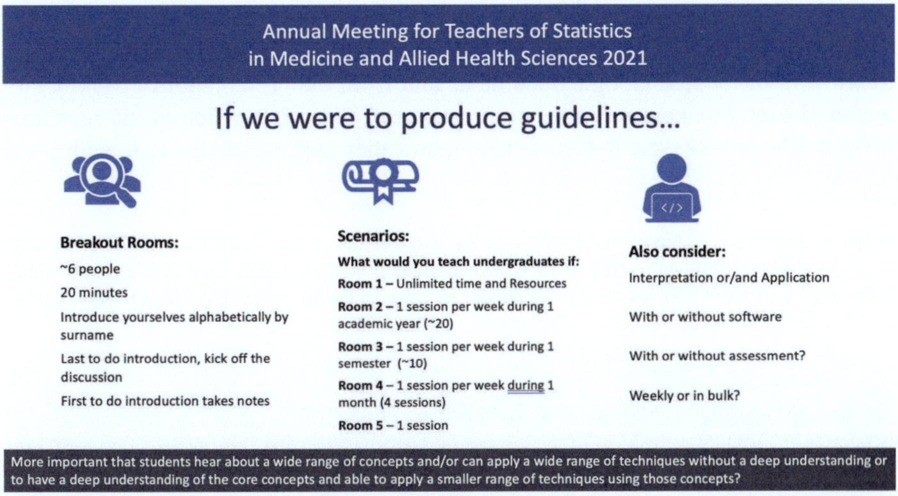

Fig. 1 Set up of the tasks provided to participants of the "Burwalls" 2021 round-table session titled: "Killing me softly with your stats teaching: how much stats is too much stats", reflecting on content and format of statistical teaching in higher education life-sciences courses in the UK

Before moving into the breakout rooms, the forum considered two aspects of teaching that were useful to frame the discussions. The first was whether we are teaching students as statistics "consumers" or "practitioners" of statistics, with one participant putting forward that students should be learning statistics as if learning to drive a car rather than learning the mechanics of the car. There was general agreement to this idea, but other participants mentioned the benefits of providing the students with an intuition of the reasoning behind statistical methods, acquired through core knowledge of concepts like inference.

The second consideration was whether it is more important to teach many different concepts and practices so that the students are aware of them, even if they don't learn or understand the principles in depth, or to teach fewer concepts with more depth — which also related to the discussion of whether we need to teach the underlying maths, or if so, how much maths is needed. In this context, it was noted that even some basic mathematical formulae such as simple divisions (e.g., calculating a standard deviation) can provide useful insights behind statistical methods. The importance of demystifying statistics was also discussed in this context, with participants sharing experiences of when they realised that some statistical concepts that they had assumed to be very complex, were indeed quite simple, and there was an element of frustration that no teachers had been able to properly break it down for them in an accessible manner.

Following the discussion, participants were asked to note on a Padlet board (https://padlet.com) which five statistical concepts/techniques would they choose to teach if that was all that they could focus on.

5 Summary of Ideas and Discussion

Ideas from the shared Google documents and from the presentations given by elements of each breakout room are summarised in Table 2 for each of the scenarios offered. The groups were not consistent in how they approached the task, with some

Table 2 Summary of the group discussions by participants of the "Burwalls" 2021 round-table session titled: "Killing me softly with your stats teaching: how much stats is too much stats", reflecting on content and format of statistical teaching in higher education life-sciences courses in the UK

Scenarios provided for time in the curriculum for statistics teaching	Main ideas from the group discussions
Room 1—Unlimited time and resources	• Most of the discussion was about the constraints currently faced as statistics teachers in higher education, which was potentially fuelled by the contrasting thought of having no constraints. As a result, the group ended up not identifying specific content that they would include for undergraduates • One of the main advantages identified by the group of working without constraints was the possibility of teaching statistics in collaboration with coordinators of other modules, to ensure there was consistency of teaching across the curriculum and that other aspects of the curriculum were not inadvertently promoting bad statistical practices and interpretations • Another opportunity identified by the group in this scenario was the possibility to spend enough time with students doing "statistics therapy", meaning breaking down of concepts by carrying out tasks step-by-step, working through different examples, listening attentively to the students' difficulties and needs through feedback, and slowly building the students' confidence in the subject
Room 2—1 session of 2 h/week during 1 academic year (~20 sessions/40 h teaching)	• The discussion focused on the obstacles faced under this scenario and on what exactly one can get across in approximately 20 sessions of statistics teaching in a year, which everyone in the group agreed was not nearly enough • The group did not come up with a shopping list of things they would include in a curriculum, but there was agreement on emphasising critical thinking (namely teaching students to think scientifically) and ensuring that the teaching is effective and relevant in terms of who the students are, and the contexts in which data analysis will be useful to them • Since many statistics teachers are not in charge of learning outcomes, the importance of working closely with people who set them was discussed to ensure consistency between outcomes and practice • Teaching the students to identify their limitations was an important consideration and it was agreed that the students should be able to know when they should consult a statistician instead of relying on the knowledge they acquire in such limited time

Table 2 (continued)

Scenarios provided for time in the curriculum for statistics teaching	Main ideas from the group discussions
Room 3—1 session of 2 h/week during 1 semester (~10 sessions/20 h teaching)	• Interpretation of statistical results and some application too, with no statistical software • Common sense approach to study design and the implications of bad design • Class exercises in data collection and sampling • Variability and averages • Relationships between variables
Room 4—1 session of 2 h/week during 1 month (4 sessions/8 h teaching)	• Dividing the sessions into one session a week for a month was agreed as the best strategy but the group was not happy with such a limited number of sessions • Considering the limited formal teaching time, it was discussed how much pre-session independent learning it is reasonable to expect the students to do, and whether this needs to be timetabled • Specific topics to be covered included: role of random chance, different types of variables, critical appraisal of research results, statistical thinking in the context of research • More emphasis on interpretation; a statistical-thinking approach • Critical appraisal of a paper: for example, start with a clinical example and give the scenario of needing to interpret a paper for a patient in an accessible manner • No software, not enough time in this course to teach an introduction to programming, but signpost them towards additional resources, such as online tutorials for programming, and other useful resources to aid learning • Recommend "The Art of Statistics: Learning from Data" as a book to follow and get examples from • Assessment would help with student engagement—including hands-on practical assignments
Room 5—1 session of 2 h	• Find a bad-quality research paper and carry out a critical appraisal, discussing strengths and weaknesses of the paper • Use the paper for breaking down the study design aspect of the sampling process • Encourage statistical thinking without getting into detail, perhaps using metaphors such as statistics help us find "signals/patterns in the noise" • Identifying ways of being an informed consumer of research • Identifying the need for statistics to answer questions with important real-world implications • Seeing statistics as a useful tool for their future career • Identifying pitfalls and good practice for their own research

following the task and questions provided for consideration and others completing the task outside that framework or debating the topics more generally without completing the task—this diversity of approaches is reflected in the summaries for each scenario, which do not follow a strict structure of ideas.

5.1 Reflections on the Challenges of Teaching Statistics in Life-Sciences Degrees

From the group discussions, it was possible to intuitively identify general themes that underline the practices and experiences of many statistics teachers in higher education. The first theme relates to the challenge of teaching a discipline that is not always given adequate time in the curriculum for meaningful teaching and learning to occur. There is an appreciation that often curricula are packed, that medical students have many other priorities in terms of their learning, and that the students are already overloaded. Nevertheless, participants shared their sense of being part of a tick-box exercise for schools to meet the guidelines and feel there is a general lack of goodwill from medical schools to address how statistics could be better represented in curricula. Some participants described having to develop courses with zero budget and no staffing time and/or having to fight to get statistics teaching appropriately timetabled to cover minimum requirements.

A second important — and related — theme relates to a sense of isolation felt by most participants in their roles as statistics teachers. This was described both in relation to their colleagues, and as general feeling of being "second-class" compared to those who teach disciplines seemingly more directly linked to the medical profession. This was also identified as a concern in relation to the students, with participants describing a greater challenge in developing relationships with the students that could aid their teaching and the students' learning, when the teachers have so little contact with the students. It was noted that this isolation of statistics as a discipline in the curriculum feeds into the students' perception of statistics as something separate and not as relevant to them as other disciplines. Another point raised related to the student–teacher relationship was some perception that the students have a strong resistance to seeking help or to admitting that they are finding the content difficult. This is potentially more common among high-achieving students such as those in medical degrees (and is probably not unique to the discipline of statistics), but it creates specific challenges for teachers who have a very limited amount of contact time with their students.

A third major theme relates to the challenges of dealing with a lack of consistency by different staff in their approach to research methods. Many participants voiced their frustration at seeing their teaching undermined by what is being actively or passively taught in other disciplines, where — for example — papers or research examples can be presented or discussed while either ignoring the centrality of statistics to research or presenting wrong or misleading interpretations of the analysis and/or results. Again, this is a problem when considering that formal statistics teaching usually occupies a very small portion of the curriculum, while the other parts of the degree can be filled with such inconsistencies.

The two groups presented with scenarios with lesser time restrictions spent more time discussing current challenges and spent less time debating teaching content or teaching format. This is interesting, as it suggests that without the time restrictions it became more relevant to discuss the wider problems of how statistics is embedded in curricula, and how consistently it is taught across these curricula, than to discuss what to teach. It can be argued that those groups with less time for teaching substantial content did not have the luxury to consider these bigger problems and were more inclined to focus on what content is essential. This is perhaps a good illustration of how many statistics teachers feel conditioned to make the most of less-than-ideal situations, instead of having the opportunity to think strategically about how their teaching could be most effective.

5.2   Reflections on Content of Statistics Courses in Life-Sciences Degrees

When content was considered, it was apparent that critical appraisal and critical skills relating to study design, sampling, and the process of inference seem to be considered the foundation of statistical teaching and the key thing to focus on if time is very limited. It is interesting to note that even the group in Room 3 with approximately 20 h of teaching in the curriculum presented a programme that focused on an understanding of sampling, study design, and simple interpretations of very basic statistical concepts, not differing much in content from the groups with more limiting time restrictions. This suggests that participants felt that more than 20 h of statistics teaching timetabled in curricula are necessary to go beyond these simple concepts. However, when asked about a choice of five core statistical concepts to teach, hypothesis testing and p-values were still chosen by a greater number of participants compared to other concepts (approximately 50% of participants). Concepts related to the process of sampling, simple descriptive statistics, and confidence intervals were also proposed by a relatively high number of participants, followed by critical appraisal and effect size/clinical significance, which were only presented by approximately 25% of participants (Fig. 2).

Comparing the outcomes of these discussion exercises with the reality observed from the survey and presented in Table 1, it is interesting to note that only 26.3% of participants reported having more than 24 h available for statistics teaching in the curriculum. In view of this, many participants often reported feeling pressure to cover more content than what they feel appropriate in the time allocated, partly to fit with expectations set from the schools, and partly to ensure the students are aware of important concepts — even if they cannot fully understand them. It is interesting to note that despite the great focus on sampling and critical appraisal during the discussions, only approximately 50% of participants reported currently teaching these concepts in their programmes, while most participants reported teaching concepts such as p-values, statistical testing, and statistical significance. Moreover, some potentially more complex statistical concepts/techniques such as ANOVA, logistic regression, and multiple regression are taught by more than 35% of respondents, suggesting that these concepts might be taught within tight time restrictions in some cases. This might be partly because some participants teach at postgraduate level where they might feel able to cover more content in less time.

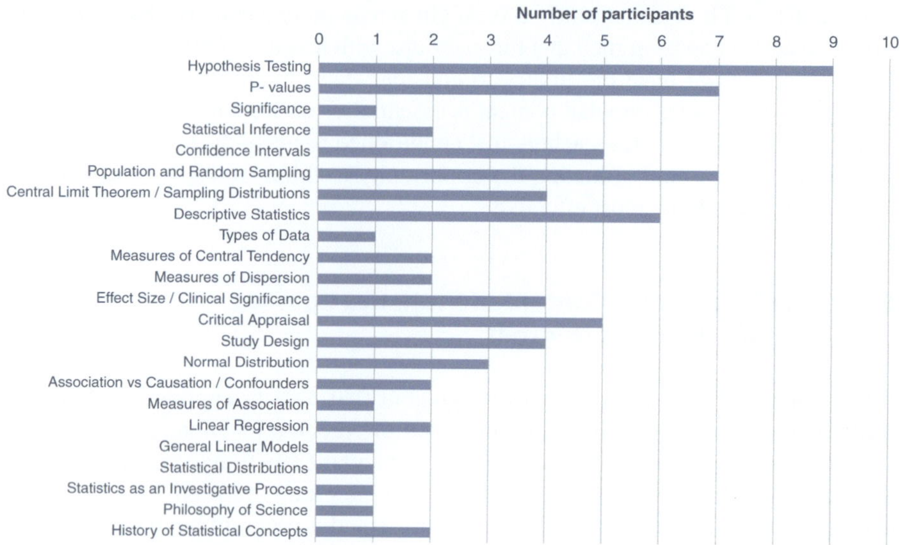

Fig. 2 Statistical concepts/techniques chosen by participants of the "Burwalls" 2021 when asked about five core concepts they would choose to teach undergraduates in the life sciences at the round-table session titled: "Killing me softly with your stats teaching: how much stats is too much stats", reflecting on content and format of statistical teaching in higher education life-sciences courses in the UK

5.3   Reflections on the Format of Statistics Courses in Life-Sciences Degrees

Regarding the question of teaching statistics with or without mathematical principles, there seems to be a divide between (a) teachers who feel strongly that this approach should be avoided as much as possible, to move students away from their anxieties about statistics if perceived as a heavily mathematical discipline, and (b) teachers who feel strongly that some basic mathematical principles are essential for demystifying of statistics, and to provide a proper understanding of statistical principles, which are themselves key to engage critically with scientific research. There was general agreement that a strong focus on complex mathematical principles was not desirable. There was also general agreement that using statistical software is desirable if there is enough time to learn it, but this can be a distraction if there is not sufficient time allocated in the curriculum.

One final point of discussion was the discrepancy between our (and a general) belief that statistics is an essential discipline in medical research, and the surprising lack of statistical competence among many successful academics and clinicians. We recognised that academics and clinicians do not need to be independent statisticians and that effective collaboration between the former and the latter is an effective way to produce good quality science. We also reflected on how the replicability crisis and some of the problems we face as teachers might stem from negative attitudes towards statistics as a discipline, and the need to involve our academic and clinical peers in a cultural shift, in order for us to be more effective teachers for our students (e.g., [14]).

6   Final Reflections and Conclusions

Forums, like Burwalls, which bring statistics teachers together, are very important for this group of people who often work in isolation and on the margins of the degree programmes they contribute to. As a network of teachers in academia, we should call for a shift in how statistics are incorporated into curricula, and how academics and clinicians across all disciplines can play an active role in such a shift. Even accepting that statistics teaching might not be given sufficient timetabled time to be properly covered in most curricula, a change in culture that emphasised the role of statistics in medical research, and ensured consistency and good practice across all disciplines, would go a long way to improve our position as statistics teachers and aid the learning process for students.

Ways to overcome the time constraints inherent to teaching statistics in medical contexts in higher education which were offered by the group include:

- Letting students know what they don't know (and won't be taught) and directing them to good resources available for them to explore these topics in their own time.
- Making the most of non-formal teaching time, such as supporting students carrying out research projects.
- Creating multiple channels to allow students to feedback and report on their needs and difficulties: requesting direct feedback during the course, including incorporating such feedback formally into the lesson plans, incorporate feedback in assessment, online feedback platforms such as Blackboard/Learning Central discussion forums, Padlets or simple survey forms, student/staff board meetings, engaging in peer review of teaching, etc.
- Using flipped learning to maximise the time with the students in the class doing practical tasks and/or discussing doubts, instead of covering content. Creating digital content for the students to explore in their own time also allows tracking of the engagement and performance data which can help planning the sessions with the students. Time considerations are important when using flipped learning because it relies on student engagement prior to teaching sessions.
- Promoting additional/supplementary asynchronous learning (especially in the context of new e-tools such as Xerte or RShiny) so that students can "learn at their leisure" in a very flexible manner, i.e., when they have time and opportunity.

Critical appraisal was recognised as a good basis and starting point for statistics learning, and one of the key skills that students should develop. It seemed clear that in medical degrees in higher education there is little expectation of teaching students enough statistics to allow them to be fully independent "practitioners" of statistics. However, engaging in practical applications of statistics can be useful for (1) developing the level of understanding necessary to be a well-informed "consumer" of statistics and (2) engaging in collaborations with statisticians to facilitate effective analyses. A greater focus on critical appraisal, as well as the reproducibility crisis and the history of statistics, might be desirable in a framework where we expect students to engage critically with research outputs; these concepts do not seem to be given much priority in current practices based on the experiences of the participants in this session although critical appraisal might have been underrepresented among this group since it does feature prominently in other, more detailed, surveys of teaching content (see chap 1: "A Review of Medical and Dental Statistics Teaching in the UK"). Future research (i.e., empirical studies) on the importance of incorporating (or not incorporating) mathematical principles when teaching statistics in medical degrees in higher education should be a priority, and more research in general on teaching and learning statistics in higher education would be very useful to guide statistics teachers and give us greater confidence in what is evidently a difficult task. Teaching evidence-based practices without sufficient evidence is clearly far from ideal!

Producing guidelines for best practice on teaching statistics in the life sciences could be useful tool to encourage those responsible for designing curricula to engage with statistics teachers in improving the conditions for teaching and learning statistics across different degree programmes in higher education. Statistics teachers should be at the forefront of this task, and we hope this chapter is a useful contribution towards achieving it. "Burwalls" is a good platform to foster further discussions and support more collaborations and publications on these topics.

References

1. Boekaerts M. The crucial role of motivation and emotion in classroom learning. In: Dumont H, Istance D, Benavides F, editors. The nature of learning: using research to inspire practice. Paris: OECD Publishing; 2010. https://doi.org/10.1787/9789264086487-6-en.
2. Gehlbach H, Robinson CD. Commentary: the foundational role of teacher-student relationships. In: Wentzel K, Ramani G, editors. Social influences on social-emotional, motivation, and cognitive outcomes in school contexts. London: Informa; 2016. p. 230–8.
3. Gillespie M. Student-teacher connection: a place of possibility. J Adv Nurs. 2005;52:211–9. https://doi.org/10.1111/j.1365-2648.2005.03581.x.
4. Hattie J. Visible learning: a synthesis of meta-analyses relating to achievement. London: Routledge; 2009.
5. Cobb G. Teaching statistics. In: Heeding the call for change: suggestions for curricular action, vol. 22. Washington: Mathematical Association of America; 1992. p. 3–43.
6. Moore DS. New pedagogy and new content: the case of statistics. Int Stat Rev. 1997;65:123–65.
7. Neumann D, Hood M, Neumann M. Using real-life data when teaching statistics: student perceptions of this strategy in an introductory statistics course. Stat Educ Res J. 2013;12:59–70. https://doi.org/10.52041/serj.v12i2.304.
8. Basturk R. The effectiveness of computer-assisted instruction in teaching introductory statistics. Educ Technol Soc. 2005;8(2):170–8.

9. GAISE College Report ASA Revision Committee. Guidelines for assessment and instruction in statistics education college report 2016. 2016. Available at http://www.amstat.org/education/gaise. Accessed 12 January 2022.

10. Aikens ML, Dolan EL. Teaching quantitative biology: goals, assessments, and resources. Mol Biol Cell. 2014;25(22):3478–81.

11. General Medical Council Outcomes for Graduates. 2018. Available https://www.gmc-uk.org/education/standards-guidance-and-curricula/standards-and-outcomes/outcomes-for-graduates/outcomes-for-graduates. Accessed 12 January 2022.

12. General Dental Council Outcomes for Graduates. 2015. Available https://www.gdc-uk.org/docs/default-source/quality-assurance/preparing-for-practice-(revised-2015).pdf?sfvrsn=81d58c49_2. Accessed 12 January 2022.

13. Miles S, Price GM, Swift L, et al. Statistics teaching in medical school: opinions of practising doctors. BMC Med Educ. 2010;10:75. https://doi.org/10.1186/1472-6920-10-75.

14. Baker M. 1,500 scientists lift the lid on reproducibility. Nature. 2016;533:452–4. https://doi.org/10.1038/533452a.

Life as a Medical Statistician

Mike Campbell

1   Introduction

For an old, retired medical statistician it is a pleasure to be invited to reflect on one's career and to give out homely advice. I'd like to thank the editors for the opportunity.

My career has been spent largely in medical schools with a few years in a research council unit so my experience is based largely on this. I have been influenced by many great statisticians, but the opinions are my own. My own career path has been extremely fortuitous. I was doing Maths at University and headed for a career in Operational Research (which I had worked in before University) when I took part in a debate and heard a pure mathematician quoting GH Hardy, to the effect that he was proud he had done nothing 'useful' and this was countered by someone else who argued on the supreme usefulness of Statistics. I promptly withdrew my application for an MSc in OR and applied for one in Probability and Statistics in Sheffield instead where I was fortunate to get a grant. My PhD in Edinburgh was on medical time series so I was ill-equipped in my first job, with the MRC Pneumoconiosis Unit in Penarth to deal with down-to-earth applications of medical statistics. I was fortunate in the help given to me by colleagues and I taught myself medical statistics by reading the first edition of the book *Statistical Methods in Medical Research* by Peter Armitage. From the MRC, I went to the University of Southampton and finally to the University of Sheffield.

Life as an academic medical statistician is extremely busy, and like all academic jobs is crudely divided into research, teaching and administration. As a rule of thumb, I think an appropriate balance is 2:2:1, although this will depend on where one is on a career trajectory. More junior people may have less administration and more teaching than senior staff. However major departures from this ratio, in my opinion, risk problems with promotion and of having less career satisfaction. For

M. Campbell (✉)
School of Health and Related Research, University of Sheffield, Sheffield, UK
e-mail: m.j.campbell@sheffield.ac.uk

© The Author(s), under exclusive license to Springer Nature Switzerland AG 2023
D. J. J. Farnell, R. Medeiros Mirra (eds.), *Teaching Biostatistics in Medicine and Allied Health Sciences*, https://doi.org/10.1007/978-3-031-26010-0_15

example, it may be difficult to get promotion with no research. I personally found one of the most satisfying aspects (and initially most terrifying) was teaching and so continued with it throughout my career.

However, numerous sub-categories come under these broad categories.

2   Research

There are four areas that seem to me to cover research as an academic statistician. The first is theoretical research, with the aim of publishing in *Biometrika*, or other theoretical journals. There is a huge amount of theoretical research in the literature, and it can be difficult to get started. Advice I got from Fred Smith in Southampton is just do what is interesting, and look up the references afterwards! However I found theoretical research difficult and time-consuming. It took me some time to write one of my more theoretical papers [1] and it has been cited by just two people! Alas, citations count in many disciplines and although some theoretical papers, such as David Cox's 1972 landmark paper on survival analysis are highly cited, many are not.

Another area is to publish applications and explanations of theoretical methods so they can be useful to others (Fig. 1). These can be published in the medical statistics literature, but if the aim is to get them used it is better to publish in the medical literature. An example of one of these was a paper I collaborated on, which garnered the largest number of citation of my career [2] (3429 citations)[1]. It contained nothing new from a statistical viewpoint but it was new to medical readers and clarified what we thought was a much misunderstood area at the time. Martin Bland once told me that the 'Statistical Notes' series that he and Doug Altman wrote in the BMJ had an h-index on their own in the highest ranking! (A h-index is easily described by an example. An author has an h-index of 50 if they have written 50 papers that are cited at least 50 times. There are also 10 year h-indices, which relate to papers only in the last 10 years). As with other measures of performance, they are of limited use, since they vary hugely by discipline.

A third area of research is to actually do research in a medical area. I will confess this has not been attractive to me. I felt that my expertise was in statistics. I did not have to treat patients and I did not understand the biochemistry, pathology, immunology, etc. required to do so. However, I have known statisticians in research groups, who become very familiar with a particular disease and 'go native'. There are also an increasing number of people who are qualified in both medicine and statistics but work mainly as statisticians in a particular area of medicine. One area where statisticians can work as primary authors is the analysis of routine data. I personally would co-opt medical specialists as co-authors to ensure I was not writing rubbish. For example, with an MSc student I found that the seasonality of asthma mortality varied by age, but I co-wrote it with doctors [3]. It has been cited often (228 citations).

---

[1] All citations Google Scholar (May 2022).

Fig. 1 Statistician's prayer: 'Lead us not into temptation'

Another area is to work as part of a team in developing a proposal, conducting the study and analysing the data. This was how I started with the MRC, and I am proud to say I have had collaborations with a number of excellent medical researchers over the years in areas as diverse as pneumoconiosis, asthma, diabetes, cot deaths and osteoporosis. In my opinion, this is the most rewarding aspect of research, and one can be imbued in the research of a particular disease without knowing all the fine details. However, it is time-consuming to build relationships and research management meetings would often take a whole day.

The way academia is set up in the UK means that doing what is personally rewarding is not necessarily good for one's career. Going for promotion, I was once asked, 'when are you going to do your own research?' I thought being an indispensable member of a team *was* my research. Of course, this is also a problem for all support staff. The person whose name is first on the grant proposal and the final paper is the one who gets the most kudos. Power comes with grant money and so even though one is part of a team, the leaders get the rewards and other members may be judged as passive.

In the UK, research is judged by the Research Excellence Framework (REF). It paradoxically encourages promiscuous workers such as statisticians to work outside one's own research group. Papers are allocated to individuals but they can only be submitted once, so it is better not to have to share a paper with someone in the same unit of assessment. Alas, expository papers such as my highest cited paper would not be submitted since they are not original research.

It is important to be aware of these pressures. Depending on how one wants one's career to go, try not to be just as the technician who is good with numbers but has no original ideas of their own. In mentoring staff, I suggest they should have at least one first-authored paper per year, which they can claim as their own, so they can counter the question that I was asked. Of course, this requires that they are given time to think for themselves. Study leave can help here, but can be difficult if a post is funded by grant money.

3 Teaching

I can recall a meeting at Burwalls when Stephen Evans from The London Hospital Medical College asked us what was the point of teaching medical statistics to medical students? After some discussion he answered his own question 'So that patients receive better care'. I still believe this!

Teaching comes under a number of guises. My view is that teaching medical students and other allied professions is what earns our keep. It is no job for the faint-hearted, teaching 300 students in one lecture in some instances. Most medical statisticians have done this at one time or another in their careers. The Burwalls Medical Statistics Conferences have been immensely useful to me over the years in showing what is possible and to be able to discuss triumphs and disasters with others.

There is an endless debate as to whether to teach statistics early or late in the medical degree. Teach it early and the students still recall their A level Maths or Physics, but usually want to get stuck into medicine. Teach it late and they are usually more appreciative, but overwhelmed by clinical work, and less in touch with numbers. Some courses now teach the bare minimum at undergraduate level which I think is a pity. It is perhaps more satisfying to teach postgraduate courses such as MScs or MPHs. Here, the students have seen for themselves how useful statistics can be and so are more motivated and the classes are generally smaller. Often also, statisticians may have to do service teaching, the finance for which usually is credited to the home department. A colleague and I did a service course once simply so we could attend the International Conference on Teaching Statistics in Brazil and present papers there.

A second aspect of teaching is basically one-to-one tuition through consultation. In Southampton the medical statistics group devoted 1 day a week to providing advice to doctors in the Health Region and kept careful logs of who we saw and where they worked. This provided useful ammunition to show the Dean, to justify our existence. This can often lead to fruitful collaboration, but as I remarked to a junior who was complaining that a researcher had not acknowledged her work 'you have to kiss a lot of frogs before you meet your prince'.

A third aspect of teaching is teaching students with a Maths background, either as undergraduates or on MScs in Statistics. I have always encouraged this as it keeps one's brain engaged with doing Mathematics, and practising this is important in the same way as doing physical exercise keeps one fit!

A fourth aspect of teaching is PhD supervision, and this also covers research. This can be very rewarding, but also immensely time-consuming if the student is not strong. One piece of advice I would give is that a thesis should be narrow and deep. Towards the end make sure that the student can give a concise answer to the inevitable first question from the examiner, 'What have you learnt from your research, and can you summarise the results in one sentence?' If the student simply describes what they have done, they need to think again.

A fifth aspect of teaching that I became heavily involved with was writing textbooks. It started with a collaboration with David Machin on a set of Tables for Sample Size calculations [4]. Since we seemed to write well together, this was followed by a concise textbook on medical statistics [5], which was based on a course we had given in China. We had to produce extensive notes for the course so it seemed obvious to write them up as a book. These books are now in their fourth and fifth editions, respectively [6, 7].

I would not advise book writing in general; the financial gain for writing is meagre and the kudos is hard to measure. It is an increasingly crowded market and I am unsure whether books are even bought by students anymore! However it was something I enjoyed and when the publishers ask for another edition it can be hard to resist. Because of 'A Common-sense approach', the British Medical Journal asked me to revise their *Statistics at Square One* book originally written by T Swinscow. It is now in its 12th edition and now at last I am the sole author and little of the original text remains [8]. Paradoxically writing chapters in books is something I would not advise, but I do because I enjoy it. (Paradoxical because this article itself is a chapter in a book).

I found teaching and writing books helped me understand the basics of our discipline. I also enjoyed engaging with younger people with their new ideas and grasp of technology, which helps you realise why you came into the area in the first place.

I was asked by the editors to comment on changes in teaching medical statistics that I noticed over the years. Obviously technology and the internet have transformed many aspects, some for the better and some not. I dislike teaching online, especially since students switch their videos off and so one has no idea if anyone is there! Much basic statistics is now taught at school, which was not the case earlier on. However, I think it is still better to start with the basics since students' experience will be varied and sometimes wrongly taught. For example, they may know how to calculate a standard deviation (sd), but will do this on unsuitable data. One pet annoyance of mine is to see sd's reported for the age of different groups, instead of something useful such as the range. (In the days when one hand wrote on the board, my handwriting must have left something to desired when a student came up after the lesson and asked why a population sd was always 6! (try handwriting σ).) The most difficult concept to teach is the *p*-value. I used to link this to the specificity of a diagnostic test, but later changed to emphasising estimation and confidence intervals over hypothesis testing. I think the most important skill to teach is critical thinking, and statistics is an excellent medium for this. I used to relate to the students Dave Sackett's quote '50% of what you are taught about treatment in medical school is wrong, we just do not know which 50%'. However it is important not to make them all professional cynics and to emphasise that good research does give reliable results.

4   Administration

For a medical statistician, administration can cover a multitude of sins.

There is the usual administration required by the University, such as to lead groups and sit on committees such as Research, Ethics or Teaching which all academics have to do and to be external examiner for other universities.

However medical statisticians are also in great demand as referees, not just for statistical papers but also for medical ones. Probably the majority of published medical papers without a statistical co-author would benefit from a statistical review! Much of this is unpaid, but requested by commercial publishing companies, who, one would have thought, could afford to pay referees. The contents of such articles are often in areas of medicine which are unfamiliar and I think it should not be undertaken too readily. The benefits of refereeing are one can see how others construct papers and get a warm glow from helping improve research, but it is often ignored in career development. However there are websites such as Kudos where one can log one's refereeing efforts. I was fortunate to be involved with BMJ reviewing early on (which was paid) with Martin Gardner, Doug Altman and David Machin. This led to our checklists of what to look out for when reviewing a paper for its statistical aspects [9]. It has recently been revised by Mohammad Mansournia [10]. Martin Gardner did a later review to see if the checklists were beneficial in terms of reducing statistical errors but found only some moderate effects [11].

There are also a huge number of other diversions which I think puts statisticians in a unique position in terms of the demand placed upon them. Statisticians are in huge demand for Data Monitoring and Ethics Committees (DMECs) since the committees are constitutionally required to have a statistician on board. However, this is no sinecure and things can easily go awry [12]. One committee I was on found that high risk infants were given the wrong dose of a drug. On another, the treatment appeared harmful and yet the stopping rules were formulated to stop early only if the treatment was effective. Often local ethics committees require statisticians as well, since as Doug Altman has said, 'Poor quality research is unethical' although Jon Nicholl (a statistician) in Sheffield has said that the main qualification to be on an Ethics Committee should be one in philosophy (his own first degree).

Statisticians are often required to evaluate research proposals, either as an external referee or on research review committees. It has been my (biased) view when I was on such committees viewing external referees that the statisticians are often able to put their finger on the nub of a problem and are often more critical than the subject matter referees, who may have their own biases in wishing to see money spent on their speciality.

I spent 10 years on a NICE Appraisal Committee, which I thoroughly enjoyed and felt privileged to be able to use statistics to make decisions about whether the NHS should prescribe a treatment [13]. It was also time-consuming (1 day per month at the committee meeting and 1 day reading the material). In retrospect it was probably far too long.

Many of these demands are unpaid, in the sense that they do not give money either to the individual or to the individual's department. There is a limit as to how much one can do in one's 'spare' time so 'real work' suffers and often so do one's colleagues if one is always away doing committee work. I was always flattered to be asked to join these committees and my predisposition was to accept requests. Although it is gratifying to be asked, in retrospect I spent too much time on these diversions, enjoyable though they may be. Be aware of the 2:2:1 ratio and if it is nearer 1:1:3 then something is askew. My advice also would be not to be on a committee for too long; on a CV what matters is that you were there, not how long you were there.

Another diversion that keeps one away from 'real work' is work for academic societies. I was Chair of the Exam Board for the Royal Statistical Society for 5 years having somehow progressed from a humble examiner (this was paid). I did sometimes wonder why I was there since I would have struggled to answer questions from some of the more advanced papers. However an important role was to keep a regular supply of examiners and I was very fortunate in getting together a good team. I was also involved in the International Society for Clinical Biostatistics (ISCB) which was a great way to meet like-minded statisticians from around the world, to meet the great and the good and get new ideas. They are often in picturesque parts of Europe, which was pleasant. It was also a good way of meeting people from one's own country! I would encourage attending conferences. One can become very insular stuck in the same place and conferences stimulate ideas, helps one to get to know colleagues and get known to people in other places. It also gives one ideas for external examiners!

In short,

1. Do a small bit of your own research, by yourself, with a team or with a student.
2. Beware of unpaid work—do some but mainly keep to what you are paid to do!
3. Join committees but not for too long.
4. Attend the occasional conference.
5. Write the occasional popular paper intending to explain new statistical methods simply to non-statisticians.
6. As a junior statistician, do not work alone, but align yourself with a medical statistics group, even just by attending their seminars.

5 Reflections

In some ways academic life has changed enormously over my career. When I started, I used to type Fortran programs on cards and simple statistical tests were done on a pocket calculator. Now the danger is that it can be too easy to run statistical tests you don't really understand. Statistics is becoming data science and some of the 'common sense' is not apparent. For example, our medical school has an 'in silico' group (i.e. simulating patients on a computer rather than using real patients). They had a project in which they were going to validate a model on as few as three patients. I will confess I tried to suggest a few (hundred) more would be better!

Much of the work that I used to enjoy doing, such as analysing data for clinicians, is now done by groups such as Clinical Trials Units and often teaching is done by non-statisticians. With regard to teaching, non-statisticians can often do a good job but I think they should be mentored by a statistician since (in my humble opinion) the emphasis can be wrong (eg always test your data for Nomality before using a t-test).

However, the tensions for a medical statistician are in many ways just the same as they were. If you are fortunate, on the whole, to do things you enjoy, you won't have regrets if it appears others have more successful careers. In some ways, the time period I was working in was a golden era of medical statistics. Papers by Nelder and Wedderburn on generalised linear models and Cox on survival led to a unification of methods and the development of computer programmes such as GLIM, MlWin, WinBUGS and R made doing advanced statistics relatively straightforward. A benefit was that the early programmes were not easy to use for non-statisticians which meant that they stayed in the hands of people who understood the methodology!

I believe medical statisticians should work in teams and meet regularly. Even if a medical department employs its own statistician, this person should be under the aegis of a medical statistics group, who can nurture and protect a lone statistician. Generally, statisticians do consulting because they genuinely want to help doctors; there are few more eager to help than the newly qualified statistician. However, I think it is highly undesirable to send junior statisticians alone into a shark-infested medical department. They need to first learn to swim in a shoal with other statisticians, where they can bounce ideas off seniors. If there are no role models available locally, I am sure senior statisticians in other universities would be delighted to mentor juniors, if they were asked.

Readers can find my (anonymous) opinions on various matters under the pseudonym Dr Fisher, in the RSS journal *Significance*, in particular *Consulting* (2004) 4: 59, *Conferences* (2006), 4: 171; *The RAE* (as the REF was then) (2007) 5: 174 and *Teaching* (2008) 5: 78.

As a junior statistician I was fortunate to be assisted by some kind, generous and able statisticians. It would be impossible to name them all, but mention must be made of Morris Walker, Peter Oldham, Geoffrey Berry, David Machin, Martin Gardner, Allan Donner and Doug Altman. Much of a successful career depends on luck and who you are working with and I think I was very lucky in many respects.

References

1. Campbell MJ. A note on the sampling of a point process for spectral analysis. J Sound Vib. 1979;62(3):471–3.
2. Matthews JNS, Altman DG, Campbell MJ, Royston JP. Analysis of serial measurements in medical research. Br Med J. 1990;300:230–5.
3. Campbell MJ, Cogman GR, Holgate ST, Johnston SL. Age specific trends in asthma mortality in England and Wales 1983-1995: results of an observational study. Br Med J. 1997;314:1439–41.
4. Machin D, Campbell MJ. Statistical tables for the design of clinical trials. Oxford: Blackwell Scientific Publications; 1987.
5. Campbell MJ, Machin D. Medical statistics: a common-sense approach. Chichester: John Wiley & Sons Ltd; 1990.
6. Machin D, Campbell MJ, Tan SB, Tan SH. Sample sizes for clinical, laboratory and epidemiological studies. 4th ed. Chichester: Wiley Blackwell; 2018.
7. Walters SJ, Campbell MJ, Machin D. Medical statistics: a textbook for the health sciences. 5th ed. Chichester: John Wiley & Sons Ltd; 2020.
8. Campbell MJ. Statistics at Square One. 12th ed. Chichester: John Wiley and Sons; 2021.
9. Gardner MJ, Machin D, Campbell MJ. Use of checklists in assessing the statistical content of medical studies. Br Med J. 1986;292:810–2.
10. Mansournia MA, Collins GS, Nielsen RO, Nazemipour M, Jewell NP, Altman DG, Campbell MJ. A CHecklist for statistical Assessment of Medical Papers (the CHAMP statement): explanation and elaboration. Br J Sports Med. 2021;55:1009. https://doi.org/10.1136/bjsports-2020-103652.
11. Gardner MJ, Bond J. An exploratory study of statistical assessment of papers published in the British Medical Journal. JAMA. 1990;263(10):1355–7.
12. Campbell MJ. Life and death decisions: a statistician on a data monitoring and ethics committee. Significance. 2005;2:116–8.
13. Campbell MJ. A statistician on a NICE committee. Significance. 2010;7:81–4.